Victoria
with special love

Adam
29ᵗʰ July 88.

UNIVERSE

I hope you
find some useful
thoughts in here.

ADAM FORD

UNIVERSE

God, Science and the Human Person

TWENTY-THIRD PUBLICATIONS

Mystic, Connecticut

North American Edition 1987
Twenty-Third Publications
P.O. Box 180
Mystic, CT 06355
(203) 536-2611

ISBN 0-89622-336-1
Library of Congress Catalog Card Number 87-50830

Cover design by Kathy Michalove
Cover photo: National Optical Astronomy Observatories shows NGC
6946, type Sc spiral galaxy in the constellation Cepheus. This is one
of the nearest galaxies beyond the Local Group. It is located 15
million light years from earth. (Use of this photo neither states nor
implies endorsement of this book by NOAO or its employees.)

To God:
'He in whose wake the year unfolds its days'
Simone Weil

CONTENTS

PREFACE

'Are you together?' I was asked by a woman in her mid-twenties at a Manhattan party. It took me a few moments to work out what the questioner meant; it became apparent when it emerged that she herself was distinctly and sadly *not* together. Fragmented relationships and emotional problems, a lack of direction in her life and a general sense of purposelessness all contributed to her unhappy state of unease. The symptoms are common. Many people in the late twentieth century suffer from a malaise of meaninglessness.

A lack of sympathy between scientific research and religious faith has partly been to blame for the spiritual disorientation that afflicts people today. Conviction in matters of faith has been displaced by intellectual doubt; people feel lost in an unfriendly world, the insignificant inhabitants of an alien universe. One version of science has reduced the world to a series of random events and mechanistic laws; in such a landscape humankind has lost its spiritual poise. But there are other ways of reading the cosmos, and science itself can be seen as the study of the world's beauty. A scientific approach to the universe, applying reason and dispelling mumbo-jumbo has immense riches to offer to religious faith; faith in turn has a light to shine on the world described by science.

My aim in this book has been to mull over some of these riches, particularly those arising from recent theories and discoveries in astronomy; the origins of the cosmos and the nature of matter, the emergence of life, the evolution and death of stars and the birth of planetary systems. Scientific research has more and more to tell us about our place in the universe. The Christian tradition complements this picture by deepening the vision and teaching that creation is rooted in

God, who is the ground of all things, from quarks to quasars, from protons to people.

The human soul is the most significant phenomenon to emerge from chemistry through evolution, but it is not a finished product. It possesses, potentially, the God-like freedom to think, speak and act creatively, yet to attain this state it is in desperate need of education, guidance and liberation. Human beings who are natural products of this universe also contain within themselves seeds for the fulfilment of a divine plan.

The Christian way contains the secret – though not exclusively, for it can be found in other traditions too – that can reveal the profound unity at the heart of all things. The evolution of life on earth may appear purposeless at one level while at another level it is full of meaning. People *can* hope to find a coherent system of spiritual truth which is part of the underlying structure of reality and not merely the projection of their own psyches. It will map out a path which is both challenging and satisfying and will lead to growth in spiritual maturity. Intellectual doubt, so much prized since the Enlightenment, will find its true place in the certainties of faith, and a vision of our role in the universe will emerge which can fulfil our deepest longings. Science and religion together have an essential part to play in this rebirth.

I am grateful to *The Times* newspaper for allowing me to explore some of these themes in its columns in articles on the Big Bang, entropy and the anthropic principle (14 January 1984, 14 April 1984 and 19 January 1985).

I am indebted too to the sharp minds of my pupils and colleagues at St. Paul's Girls' School and to the searching questions of my wife Veronica, all of which have helped me to think.

1. SCIENCE AND MYSTERY

'Scientific research is simply a form of religious contemplation' wrote Simone Weil,[1] French philosopher and mystic. It is a paradoxical thought; modern warfare's ability to blast, burn and radiate millions – a direct result of science – seems to confirm the alternative view that science is the enemy of religion. But all new-found freedoms contain good and evil options, and knowledge is a risky commodity; the more we know the more power we have to create or destroy. For all the inhumane uses to which science may be put, it can also open our eyes to the God-given reality that surrounds us and can reawaken that vision of the universe perceived by the shepherd Amos in the eighth century BC:

> 'It is he who made the Pleiades and Orion
> who turns the dawn to dusk
> and day to darkest night . . .
> Yahweh is his name.' (Amos 5:8, JB)

This insight, that creation is an expression of God's will, is becoming, for many people, more rather than less true as science advances, uncovering the nature of reality from sub-atomic particles to the structure and evolution of vast galactic star systems. In the late twentieth century mystery is being rediscovered through scientific research.

The practising scientist nowadays may find it as reasonable to believe in God and to pray as it is to probe the atom or to wrestle with a mathematical theorem. The days are gone when it was assumed that the pursuit of scientific method ruled out the possibility of holding religious beliefs. These two areas of thought, once treated separately, now contribute jointly to a coherent and integrated vision of the world. Science and religion offer insights into one reality and together can help build a framework of faith that encompasses the whole of contemporary life.

Lesslie Newbigin in a small and challenging book addressed to the British Council of Churches, *The other side of 1984*,[2] has appealed for Christianity to take a step beyond the discipline of doubt imposed by the Enlightenment. We need, he argues, to return to dogma. Not the blind dogma of creeds repeated by rote, respected simply because they are traditional, but the dogma of believers who are sure in their faith. This dogma will require a new framework of faith. It is a framework which will be as true to the discoveries of science as it is to the revelations of theology.

The greatest threat to people today is the sense of meaninglessness. Viewed one way the world described by science can seem terribly cold and haphazard. The haunting fear is that religion offers only illusions to protect us from the truth that the universe is ultimately a stark, uncaring place. Blind laws irrevocably work out their aimless patterns and we are no more than one of the chemical accidents of nature. Not only are we purposeless like everything else but we are cursed with the capacity to know how purposeless we are. All our attempts to find meaning are simply a desperate effort to make reality more comfortable; but artificial myths and invented gods eventually wear thin.

This spiritual doubt poses a seismic threat to humankind's health and happiness. Perhaps more than at any time in human history people need to be positive about the meaning of their lives and to know where they are going. We have in our hands the power to alter life itself through genetic engineering and scientists may shape the world of the future. We are already transforming our environment and possibly pushing the world's climate in a new direction. Deep in our psyches, since the dropping of the atomic bomb on Hiroshima on 6 August 1945, there is the realistic fear of a nuclear holocaust in a few moments of madness. In the face of these immense moral and spiritual dilemmas, we desperately need guidance and to be reconnected to a spiritual reality from which we have become alienated. Nostalgia for this dimension to life is abundantly apparent in a renewed interest in the world's mythology, in the developments of Jungian psychoanalysis, and in the proliferation of new religions and sects.

The hopeful news however is that a new sense of humility is abroad, arising from the discoveries of science itself. The sense of wonder increases rather than diminishes as we begin to make coherent sense of our place in the universe. Science's insights into reality can generate a new sense of mystery based upon knowledge rather than upon ignorance, and science has the capacity to lead us back to a new sense of awe. Ever since the Enlightenment there has been an entirely proper urge to doubt, to question, and to rationalise. Hard thinking and rationality are a necessary part of the quest for understanding, but they are tools and have their limits. Not all that is good, true or beautiful can be captured by the neat networks of the human mind. The experience of awe may help us take the not unreasonable step beyond reason. However enlightened we may be intellectually or scientifically, there is still a need to stand and tremble before the holy, to take off our shoes and be silent. This is every bit as true for the theologian as it is for the scientist.

The word 'mystery', although still devalued and debased in some quarters, is beginning to enjoy a renaissance, and this rebirth is very much the result of progress being made in science. Two different uses of the word should be distinguished, for their confusion has led to the misleading idea that science is the enemy of mystery. This assumption had become well lodged amongst the prejudices of the twentieth century. In everyday usage mysteries are often no more than riddles, problems or conundrums to which a solution has not yet been found. The plot of a detective story contains such a mystery; 'whodunnit' will be revealed on the last page and the mystery will disappear.

It is the task of science to unravel such mysteries in the natural world, to explain and dispel them. The analogy is often used of a torch lighting up dark corners, expelling ignorance, fear or superstition and bringing clarity. Why are there thunderstorms? How do plagues spread? How did the world begin? How did life first emerge from the chemistry of 'dead' molecules or how did such a complex organ as the human eye evolve by natural selection? How did the elephant get its trunk or the bumble bee fly with such aerodynamically inadequate wings? How do newly hatched whitethroats manage to migrate successfully to north Africa or a flock of ten thousand starlings turn in unison? How does consciousness relate to brain chemistry? Why do we feel guilty? Why does cancer strike some people and not others? As scientific research makes headway many of these intriguing or pressing problems yield to solutions. The systematic use of reason coupled with observation is followed by explanation and understanding as the intellect makes sense of the world and mysteries are put to flight.

For some people this role of science as the dispeller of mystery has been allowed to step beyond its proper limits of problem solving into a totally different realm. Not appreciating that a distinction has to be made they imagine naively that science will one day dispel *all* mystery.

There is another more profound sort of mystery, the subject and object of religious faith, which lies beyond, though not contrary to, the use of reason. The proper human response is one of wonder and awe, not simply an attempt to resolve the mystery as though it were an intellectual or scientific problem. There is more to 'knowing' than rational explanations can give of any particular phenomenon, whether it be a poem, a piece of music or a person. The exercise of reason is unconditionally necessary in all our accounts of the world, but it is not in itself sufficient. When a rationalist has explained the mechanism of evolution in terms of Neo-Darwinian theory an all-embracing mystery is still there; when neurophysiology reveals the material basis of thought mystery does not disappear; when Grand Unified theories link all the forces of nature and expose the mathematical basis of creation and the elegantly simple origins of all the richness of the cosmos mystery is not dispelled but enthroned.

One of the great discoveries of the twentieth century is that science itself has a major role to play in feeding religious faith by awakening a sense of awe. 'The mystery (or secret) of the kingdom of God' proclaimed by Jesus Christ (Mk. 4:11) has a material aspect which an investigative, rational approach to the world is uncovering. Clarity of thinking in this remarkable universe can only open the mind and heart to wonder. As mysteries of a lower order are dispelled by scientific research, higher order mystery can really come into its own.

Research into the fundamental properties of matter at places like CERN (the European Council for Nuclear Research) near Geneva is now contributing to a religious vision of the universe. The super proton synchroton at CERN is a seven kilometre ring of pipe flanked by giant magnets which runs beneath the cornfields and vineyards of France and Switzerland. Massive pieces of equipment needed to detect the fleeting debris of particle collisions have to be housed in a building the size of a cathedral. The picture of the basic structure of matter which emerges from theory and research is quite remarkable and awe-inspiring.

The building brick of people, mists, and mountains is the atom. Until this century these minute objects were thought to be like midget billiard balls obeying the laws of classical physics. They were thought to be 'uncuttable' as their name from the Greek suggests. Now we know that each atom has a remarkable and subtle inner structure. In some ways it is literally beyond our imagination. Since 1911 when Rutherford revealed that the atom contained a nucleus orbited by electrons and proposed that it was analogous to the solar system, the atom has revealed even deeper layers of inner complexity. Fundamental particle physicists are still unwrapping the layers of this 'cosmic onion'. The nucleus, it turned out, is composed of varying numbers of protons and neutrons (depending on the element) and now it is revealed that these subatomic particles are themselves composed of quarks with exotically named qualities like charm or strangeness. But what is truly extraordinary is that all of these subatomic particles are themselves simply 'frozen' manifestations of pure energy. Quantum mechanics, part of the new physics, offers an even weirder description of the atom. All of these fundamental particles may equally well be described as waves. Now a wave is a very different sort of phenomenon from a particle; yet – depending on the experiment – an electron, for instance, may appear as either. This wave particle duality puts an impossible strain on our mental capacity and we simply have to accept that it is the case. A wave is an undulation in a substance while a particle is a packet (hence the word 'quantum' of quantum mechanics, meaning a 'packet' of energy). Reality really is beyond our imagination.

According to Niels Bohr quantum mechanics has even more mysterious things to say about the bundles of energy (waves or particles) that make up the atom. It seems they are nowhere in particular until the scientist observes them. Each is simply a bundle of possibilities, a host of ghosts of what *might* be. When the observer makes an observation then one of these ghosts rises to the occasion and becomes real. The act of looking brings a clarity to the particles that they do not have in themselves. Can anything be more ultimately bizarre than that? A human being by making an observation seems to take on a creative role and focuses the ghostly fuzziness of atoms into sharp reality. What they are when we do not look at them is quite beyond our ken.

There is an awful lot of emptiness in an atom. The ghostly possibilities described by quantum mechanics haunt a near void. The nucleus of protons and neutrons is surrounded by shells of energy created by electrons orbiting this nucleus millions of times per second. Rutherford's tentative solar system analogy attempts to describe the scene. The space between the electrons and nucleus however is proportionally speaking comparable to the emptiness of St. Paul's Cathedral. If we represent an atom by such an edifice then its nucleus is no bigger than a fly buzzing somewhere halfway down the nave. It occupies a volume less than a thousand millionth of the space of the whole atom; almost a void, yet packed with power. (Remember it was by releasing the energy pent up in this nucleus that the city of Hiroshima was wiped away in seconds.) If it is a carbon atom we are modelling then its six electrons would have to be represented by tiny specks of dust distributing themselves at unbelievable speeds between the dome, the high altar and the great west door.

Remove the inner space from an atom and pack all the atomic nuclei together and the matter contained in the world's largest building would fit easily on the head of a pin, (though not for long, the mass being undiminished). Such extraordinary conditions *do* exist in neutron stars where the force of gravity has caused the space in the atoms to be 'squeezed' out of them as it were, while the nuclei are jammed together. A handful of such a star would weigh many millions of tons. Even more astonishing, if all the matter that makes up Planet Earth could be pulled together by gravity into the sort of compactness that exists in a 'black hole' (exotic stellar objects for which there is an increasing body of evidence) then it would occupy a volume of space less than two centimetres in diameter. Packed that way, our world, its core and crust and continents, its oceans and mountain ranges, its forests, cities and people would be reduced to the size of a child's marble.

As we knock our way around this hard everyday world, we live, it seems, within something of an illusion. Even steel girders and mountains are no more than a gossamer of energy. In a telling phrase the physicist Paul Davies has said that we and all other material things have been 'spun from a frolic of Nothingness'.[3] All hard things, which seem so solid to us – tables, paving stones and heads – are in fact like fine three-dimensional lace, a sort of ghostly spider's web woven from energy as energy performs its unending cosmic dance.

The language of the new physics begins to sound like mysticism and several writers have begun to draw parallels between eastern traditions such as Taoism or Zen and the philosophical speculations of some physicists (e.g. F. Capra in *The Tao of Physics. An exploration of the parallels between Modern Physics and Eastern Mysticism*; and G. Zukav in *The Dancing Wu Li Masters. An overview of the New Physics*). The convergence of eastern mysticism with the new physics can be overstated, however, and usually takes the form of generalisations, vague analogy, and references to experiences in meditation of the interconnected unity of all things. The value of these exercises in drawing parallels (apart from giving rise to very entertaining and thought provoking reading) is that they encourage western minds to see that there are more ways of speaking about the world than simply with the descriptive rational language of science.

The current interest in eastern mysticism might be no more than a passing fashion, a fad, were it not for the fact that the west has its own mystical tradition which usually running underground is beginning once more to come to the surface. The appeal of Zen, Taoism and Indian Vedanta stems partly from the western malaise of alienation. A scientific approach to the world has temporarily detached us *from* the world. But the mood is changing. Both western spirituality and western attitudes to nature are more concrete than in eastern traditions; four hundred years of science has developed in us an understanding of physical reality not available to the great spiritual giants of the east, Buddha, Shankara, Lao-tzu, Bodhidharma. We know more about the ground we stand on. Western mysticism is more in tune with the unyielding solidities of a physical universe and when it is rediscovered it will heal the fragmented view of reality produced so far by science, and will have a much more powerful spiritual way to offer than comes to us from the east. The west having plunged more deeply into the reality of matter will achieve a greater unity in the end. Western science is taking us beyond the mysticism of the east.

The world of the atom is not the only area in which science reveals mystery. At the other extreme from the microscopic nothingness of particle physics, the macroscopic universe of galaxies is explored by astrophysicists. Astronomy, that branch of science which with mathematics first dislodged humanity from its central place in the universe, is itself offering new insights for religious faith. Copernicus in the sixteenth century showed that the earth and planets orbit the sun; he seemed to move humankind from the focus of creation. Today astrophysics helps to reinstate humanity by showing that it has profound links with the stars. People are not merely insignificant dust particles in an alien cosmos. Astronomers, working with x-ray and infra red detectors orbiting on satellites above the obscuring effects of our atmosphere or listening in to deep space with radio telescopes to events that happened billions of years ago, have made startling discoveries which provide new insights into our connection with the stars.

Astronomical observatories spread like crops of squat white mushrooms around the peaks of some of the most exotic mountains in the world. Several of the largest infra red telescopes are sited on the top of an extinct volcano Mauna Kea in Hawaii in the Pacific; the 100-inch Isaac Newton telescope has been moved from southern England to a mid-oceanic mountain site, La Palma in the Canary Islands, to gain improved seeing conditions. The largest solar telescope of its kind sits on the top of Kitt Peak, Arizona, a mountain held to be sacred by the Hopi Indians. From these and other Olympian sites astronomers probe deep into the universe and back into time. The vastness they reveal can induce a sort of spiritual vertigo, and yet a new thought is dawning. It is a vastness to which the emergence of life on earth and humankind, including the astronomer with telescope, is intimately and intrinsically related. If the universe were otherwise we probably would not be here. The atoms we are made from, the stuff of our flesh and blood, originally came to earth from the stars so that in a metaphorical sense we are the eyes of the universe. The cosmos is not a cold unfriendly place somewhere far 'out there'. Quite the reverse. Some physicists today have a hunch that the basic laws of the universe were designed to work together to produce observers with minds.

In humankind the universe wakes up and finds its meaning; in us the cosmos converses with itself.

Science highlights mystery, but no particular area of research, whether at the macroscopic level or the microscopic, will ever reveal by itself whether there is a God or not. In this I agree with the atheist. At the scientific level of description the universe is a self-explanatory system. When I look through my telescope down on to the crumbling mountains of the moon bathed in the white light of the sun or at the everchanging ammonia and methane clouds of Jupiter, or when I peer further out into space at a globular cluster of half a million stars or a neighbouring galaxy I see nothing which tells me directly whether I or the atheist are right in our different beliefs. The same is true for any other area of research. Science, to the end of time and to the limits of space (if such phrases have any meaning), will build up an autonomous network of knowledge. Everything from the first billionth of a microsecond after the Big Bang when this universe was born, up to the evolution of humankind will eventually have an adequate and satisfying explanation. Even the apparently intractable problems such as how life emerged from chemistry or minds emerged from mammals will finally yield solutions. Faith will no longer be able to smuggle God in to account for these dark corners in the system of explanation we erect to describe the world.

Between the vast dimensions of cosmic space and the microscopic world of the atom there is another mystery-evoking realm, the evolution of life. Life and mind are the most remarkable emergent qualities of matter, and in the human person—where chemistry is at its most complex—a new phenomenon appears on the face of the earth. A coherent account can be given of a human being in terms of chromosomes, biochemistry, and environmental conditioning. But to reduce a person to chemistry and then to argue that he or she is nothing but a sophisticated chemical engine is to perpetrate one of the most unfortunate fallacies of the twentieth century. The reductionist fallacy makes the mistake of confusing the quite proper methodology of science, the careful analysis of any given phenomenon into its constituent parts, with the philosophical judgment that there is no more to the phenomenon *than* those parts. The reductionist sticks too close to the details without seeing the significance of the whole picture, rather like someone examining a newspaper photograph with a microscope and seeing only dots.

Christian faith will be committed to the view that 'emergent qualities' such as life and mind, although not present in the individual atoms or molecules of an organism are nonetheless just as significant as the material base from which they emerge. A holist approach, facing the opposite way from the reductionist, views the whole functioning system of a person as a unit. When we then ask questions about the human spirit or about beauty and ugliness, good and evil, justice, truth, and value we are exploring fields just as real as those researched by particle physicists or chemists. When a human being has been described in accurate physical detail by science, the functioning of every molecule in the brain accounted for by neurophysiology mystery still remains. It is even enhanced. To deny it would be as blind as to argue that T. S. Eliot's *Four Quartets* are merely a complex arrangement of the letters of the alphabet.

Human beings, according to biology-watcher and essayist Lewis Thomas 'are the newest, the youngest and the brightest things around'.[4] It is here that theology in the twentieth century will concentrate its focus, for with human beings emerges meaning. If, as Christian faith suspects, the universe is grounded in spirit then material phenomena are only one aspect of reality. The new physics has acclimatised us to the thought that space and time are two aspects of one interconnected system, space-time, and that matter and energy are interchangeable according to the formula $E = mc^2$ discovered by A. Einstein. Quantum mechanics presents us with the mind-warping paradox that all fundamental particles can be conceived as particles or waves. In such a space-time cosmos it is not unreasonable for faith to propose that physical and spiritual realities are two aspects of one deeper reality. The universe is not just the material system investigated by science – it is an evolving process of spirit-matter.

Awe is one of the most ancient emotions in religion. Our earliest written records show men and women filled with fascination and terror by high places, deep pools, the dark or thunderstorms. With greater spiritual sophistication the emotion was transferred to more abstract ideas, the glory of God or the oneness of the creator. But awe is not enough. As James Watson, co-discoverer with Francis Crick of the helical structure of the DNA molecule, put it succinctly when lecturing at St. Paul's Girls' School, 'you don't have to believe in God to be awe-inspired by the DNA molecule'. Mystery is here to stay and science is its servant every bit as much as theology. But awe is just the beginning of wisdom. As the Old Testament reminds us, holy fear is the start of a spiritual journey not the end. It is not enough simply to be awe-inspired. Awe, which is the basis of faith for many people, contains an invitation to respond. We are born from a spirit-matter process and have a vocation to be human which involves us in developing a relationship with that mystery which is the source and ground of the universe. Awe makes demands.

Christian thinkers will explore the mystery of the human phenomenon with Jesus Christ as the guiding figure. In Christ the full potential of humanity is revealed. By the incarnation the deep unity of spirit and matter is demonstrated and through this revelation the ultimate purpose of the universe uncovered. The judgement of St. Athanasius in the fourth century AD remains true for us today, that in Christ 'God became man in order that we might become God'.[5] Jesus is the meaning made flesh. Evolved from the earth, humanity has a remarkable destiny, should it be awake to the opportunity. It is a receptacle for grace, a focus for the Holy Spirit, capable of sharing with God in the act of creation.

Scientists and theologians are now well set to work together as interpreters of reality and to discover that it is richer than anything they can say about it. Science, the dispeller of mysteries, becomes the proclaimer of mystery while theology reveals meaning. The faith of a Christian today can be fed as much from the pages of *New Scientist* as from the New Testament.

2. THE SUN AND STARS

The finest patch of night sky, to my mind, can be seen high overhead, from British latitudes, on a clear winter's evening. Some of the brightest stars shine from the constellations of Orion, Taurus, Gemini, Canis Major and Minor.[1] One needs to get away from the lights and polluted air of a town to view them at their best; an open space with a low horizon to the south. Orion the Hunter stands well above the hedgerows, the first magnitude star Betelgeuse burning dull red in his right shoulder (or left as we face him). A straight line of equally spaced and easily identifiable stars define his belt while another brilliant first magnitude star Rigel shines with a green white light from his left foot.

Orion, straddling the horizon, is one of the few constellations to look at all like the figure it is supposed to represent. As the night progresses the hunter swings from east to west in the heavens as the earth slowly turns beneath the sky. Following him, but nearer the horizon, is Sirius (from the Greek for scorching) the Dog star, the brightest star in the whole sky. On a frosty night it may seem to flash red, green and white as the light rays (after years of flight through empty space) are refracted and disturbed by our turbulent atmosphere. Sirius is a close neighbour in space, in astronomical terms. It has a companion star not visible to the naked eye and together they belong to a local group of stars of which our own sun is an unspectacular member, a 'yellow dwarf' to be precise. It is only the sun's proximity to earth that makes it outshine all other stars. It dominates our days with its blazing heat flooding the atmosphere with light and turning the sky blue, but now since the invention of the telescope it has had to take its proper place amongst the stars as astronomy reveals their true nature. Whereas it takes light only eight and a half minutes to travel the 93,000,000 miles from the sun to the earth it takes eight and a half years to arrive from Sirius, a sun twice its diameter but half a million times its distance. Light, travelling at 186,000 miles per second, covers about fifty million million miles in eight and a half years and so for convenience Sirius is said to be eight and a half light years away. Only half a dozen stellar objects are nearer to us than that. When exploring and contemplating the universe we find that even local distances are vast and almost beyond imagining.

The stars in the constellation of Orion, though seemingly close to Sirius in the sky, lie far beyond. Betelgeuse a red hot ball of gas called a red giant is two hundred million miles in diameter and 470 light years distant. The light we see today left the surface of the star in the days when Henry the Eighth was a young boy. We see the star as it was then in the fifteenth century. Rigel is even deeper in space and it takes 1,300 years for its light to get here. Rigel is truly a giant sun, for it shines some fifty thousand times more powerfully than our own sun which explains why we see it twinkling brightly from this immense distance. The starlight that registers on the retinas of our eyes on some winter evening left that brilliant sun in the dark ages of the seventh century two hundred years before Alfred the Great became king. If Rigel were to explode today it would be thirteen hundred years before the event is witnessed on earth. Yet when we think about stars like Betelgeuse or Rigel which are visible with the naked eye we are still only contemplating our closest neighbours out of the hundred billion stars that make up the 'island universe' that is our galaxy. And the galaxy itself is but one swarm of suns in a cosmos which contains billions of galaxies.

The true dimensions of space only began to dawn on astronomers in the eighteenth century. Using Kepler's laws of planetary motion they were able to calculate reasonably accurately the distance of the sun from earth. One writer then made the somewhat romantic estimate that it would take 1,500 years to reach the sun on horseback! In practice a scale model of the solar system would be unwieldy and one of the galaxy impossible, so much empty space is involved. If for a model of the solar system we were to choose a brick-sized ball to represent earth then the sun would have to be as big as a house a whole kilometre away. The distances involved in the galaxy present an even greater problem. Represent the sun with a grain of sand and we would have to travel three kilometres to drop the next grain representing the nearest star. Planet Earth suspended in this vast void shrinks to virtual vanishing point.

The creation myth of Genesis tells us, almost as an afterthought, that God 'made the stars also'. The authors of Genesis although inspired by God to write down their convictions about the creator and a human's place in creation, could not have known that the tiny pinpricks of light in the night sky which are so powerfully outshone by the sun and moon, are in fact vast nuclear furnaces of cosmic dimensions and immense antiquity.

For thousands of years the 'peace of the shining stars', to quote a prayer from the windswept Scottish Isle of Iona, has brought calm and stillness to the person who contemplated them. The quiet constellations swing high above the turbulent weather and the storm clouds of human anxieties. That peace is in danger of disruption by the new flood of knowledge being made available to us in the twentieth century. A misreading of this knowledge can hinder our understanding of ourselves and our place in the universe.

Nowadays the stars make us feel small. Astronomy has made us peripheral and reduced us to no more than infinitesimal dust in the corner of a frighteningly vast cosmos. Our culture has not yet absorbed and properly digested the new perspectives provided by scientific research and we have not adjusted to what we now know about the stars. Those gigantic foreign suns deep in space are so far off that we become quite giddy trying to comprehend the distances. Sigmund Freud voiced the effect this has had upon our souls. 'Humanity has in the course of time had to endure from the hands of science two great outrages upon its naive self-love. The first was when it realised that our earth was not the centre of the universe but only a speck in a world system of a magnitude hardly conceivable. The second was when biological research robbed [us] of [our] particular privilege of having been specially created and relegated [us] to descent from the animal world.'[2] Neither of these blows need be as undermining to our self-esteem as Freud implies. I will examine the apparent, but I believe superficial, threat that biological research poses to our status in nature, in a later chapter. For the moment my concern is to explore what astronomy has to offer in our search for meaning. The universe is yielding up many of its secrets in our time and not all of them need make us feel small. Quite the reverse in fact. Many of the facts uncovered by astrophysics can contribute to, rather than devalue, our sense of the mystery of the human individual. Much depends on how we ponder them.

Not only is the universe vast but it is also ancient. The ten thousand years of human civilisation from the foundation of the first settled villages at the end of the last ice age up to the computer revolution of the twentieth century is nothing compared to the life cycle of a slowly turning galaxy. One gyration of a galactic star system such as the Milky Way takes 260 million years. But from the viewpoint of our transitory lives celestial movement seems frozen. Everything in the universe is in motion but the distances are so great that time seems to stand still amongst the stars like the hour hand of a clock. Consider for example the Hyades. The Hyades are a 'V' shaped cluster of stars which form the head of Taurus the Bull. Taurus is above the left shoulder of Orion (the right as we view him). The eye of the bull is the red giant Aldebaran, the brightest star in the constellation, and the angle of the head gives the impression that Taurus is charging at the Hunter Orion. In fact we now know from accurate measurements on photographic plates that the whole group of stars are members of one cluster of some 132 suns and they are winging their way through space straight towards Orion's head at forty kilometres per second. But they are so far off you won't see them move as you watch. Even if you could go to sleep for ten thousand years and then look again you would notice very little displacement with the naked eye. On the cosmic clock human life is but a few seconds, like those butterflies which are here today but gone tomorrow. Buddhist philosophers came closer to appreciating the age of the universe than their western counterparts. They compared an aeon, an age, to the time it takes to erase the world's highest mountain by dusting it once a century with a silk cloth. The earth is no exception to the discovery that everything in the sky is moving. The psalmist for all his wisdom got it wrong about the fixed earth set on foundations that cannot be moved. 'He laid the foundations of the earth' he says of the creator 'that it never should move at any time' (Ps 104:5). In fact we are floating free, adrift in space and the drift is far more dramatic than that caused by any current in the wide oceans of the earth. Copernican mathematics set us loose to orbit the sun; modern astrophysics has given us the wings of Mercury. The earth flies silently around the sun, our parent star, at thirty kilometres per second. But the sun itself is not a fixed point in the universe standing still amongst the stars. It is drifting with its attendant planets at thirty kilometres per second towards Vega in the constellation of Lyra the Lyre. Meanwhile the whole galactic swarm of suns to which we belong, the Milky Way, is itself tumbling through intergalactic space at 600 kilometres per second towards a remote cluster of galaxies far beyond the constellation of Virgo. There are other motions that we have to take account of too, such as the slow spin of the galaxy about its centre like a giant Catherine wheel.

In the time it takes to say 'a thousand kilometres' we have moved a thousand kilometres. We were space travellers, we find, even before we sent men to the moon. The bed you climb into tonight has moved many millions of kilometres from where it was in the universe when you got up this morning. Earth is a tiny life raft in a cosmic ocean and the security the psalmist found in contemplating a fixed earth we must find, but at a deeper level. As denizens of the cosmos we fly with our planet; as children of God we may perhaps find rest in stillness.

All this vastness, all these aeons, have undoubtedly influenced the way we think of ourselves. Twentieth-century people are threatened by a deep sense of meaninglessness; ultimately, nothing seems to matter. These troublesome doubts however are the result of a misreading of science, a superficial assimilation of the facts of astronomy. The implications of the new perspectives offered by astronomy can, however, show humankind in an entirely new light in which the significance of the human individual acquires a new status. If we contemplate no further than first impressions then we are rather like the astronomers of Plato's *Timaeus*. Plato propounded a view that the world was a degenerating copy of an ideal world which existed in the beginning. It involved a sort of evolution in reverse whereby people in subsequent incarnations became animals. 'Birds' he wrote 'were produced by a process of transformation, growing feathers instead of hair, from harmless empty-headed men, who were interested in the heavens but were silly enough to think that visible evidence is all the foundation astronomy needs.'[3] A prevalent and similar fallacy of the twentieth century is to make value judgements on the basis of size, and to think that the phenomenon of the human mind can be measured against the dimensions of space.

Without the sun and the stars we would never have been here. Let us begin with our immediate parent the sun, that glorious orb which floods the skies with light, turning them blue. The sun has always had a central role in the mythologies of the world. It was the winged god Shamesh of the Babylonians who looked down on all human activities and handed to Hammurabi a law code to guide them. He was Amon-Ra and Aton to the Egyptians. Some of the hymns that praised him as Aton have filtered into the Bible, according to some authorities, as psalms in which the glorious imagery of the blazing solar disc is transferred analogically to Yahweh (cf. Ps. 104). In Europe he became most famous as Mithras, originally a Persian God but later popular with

Roman soldiers. Mithras' birthday, 25 December, was hijacked by the Christian church and rededicated to the birth of a new morning star, Jesus Christ, and the dawn of a new era.

Jewish priests in Babylonian exile in the sixth century BC were careful to avoid any implication of sun worship when they told their story of creation in Genesis. The sun is demythologised in that piece of poetry. They avoided use of the Hebrew word 'shamesh' since it was the name of the God in Babylonia where they were captives. The sun becomes merely an instrument of God created on the fourth day to give daytime to the world, a light in the sky and no more.

The sun remained little more than a benevolent light in the sky up until the sixteenth century. The world was a small stage at the centre of the cosmos; daytime gave people the opportunity to go about their business while nights were designed expressly for rest from all their labour.[4] Even so, an informed suspicion was abroad, and had been since the days of the Greek philosophers, that there is more to the sun than meets the eye. By the second century BC the Greeks had calculated that the sun's distance must be at least 140 times 4,000 miles. It had to be an immense object. They knew the curvature of the earth by the sun's angle above the horizon at noon from different cities. They were then able to calculate the earth's radius as roughly 4,000 miles. It followed that when the sun is just above the horizon at dawn or sunset it must be 4,000 miles further away than when it is directly overhead. It ought then to appear larger at noon than it does at dawn, that is if it lies reasonably close to earth. Since their instruments could detect no increase in size whatsoever when the sun reached the zenith they reckoned that it must be at least half a million miles away and therefore gigantic.

We now know that the sun is a vast nuclear furnace some ninety-three million miles from earth. We owe everything to it; our planet, the evolution of life and all of our everyday energy needs. Within hours life on earth would cease if our parent the sun were to vanish. Every second the sun turns four million tons of mass into energy as the nuclear processes deep in its interior convert hydrogen to helium. But in five billion years this has made very little inroad into its total mass.

Current astronomical theory suggests that God needed the sun before God could make the dry land appear, unlike the account in Genesis. Five billion years ago a tenuous cloud of dust and gas in interstellar space underwent a process of gravitational collapse triggered perhaps by the shock waves from an exploding star known as a supernova. At the heart of the cloud the material was pulled together so densely that nuclear fusion reactions began and a star was born, lighting up the remnants of the cloud around it. There are several theories about how planets were then created in orbit around the parent body. They invoke magnetic fields as well as gravitational forces and speculate about whether the discs of dust and gas that condensed gravitationally into planets were left over after the formation of the sun or were breathed out from the solar furnace when the protostar was still unstable. Whichever scenario is nearest the truth it remains a fact that without the sun the material that came to form Planet Earth would still be a nebulous dark cloud in empty space, a million times less substantial than domestic dust dancing in a sunbeam. Without the sun there would be no world.

A world without a steady source of energy to keep it at the right temperature would never have evolved life. On earth the conditions have been just right for the past four billion years and it has taken that long to evolve intelligent creatures. As far as we know it is the only planet in the solar system to support organisms of any sort. Had the earth been only one and a half percent closer to the sun then its atmosphere would have suffered a runaway greenhouse effect with surface temperatures approaching 400° centigrade day and night as they do on Venus. If, on the other hand, the earth had been one and a half percent further away from the sun then it would be frozen in a permanent ice age rather like on Mars.

The energy that floods down on us daily is only a minute percentage of the total output of the sun and negligible compared to the amount which radiates out into space. But even on the night side of earth we do not escape the sun's shining. One of the products of nuclear fusion is the neutrino, a chargeless massless particle. These subatomic entities are so unaffected by other matter that they can fly right through the earth and pass out the other side. As you lie in bed at night you sunbathe like a ghost as billions of neutrinos from the sun flash up through the earth through your body and on out into space. From the sun's point of view the earth might almost not be here.

One day the sun, upon which we are totally dependent, will die. At the present it is a Main Sequence star, which means that it is 'burning' stably and steadily and will do so for another five billion years; it is only half way through its career. Humankind has a bright future provided it manages to survive the risks of adolescence. But in a remote epoch the sun will undergo a radical transformation and the descendants of the human race will have to migrate. The sun will expand from the yellow dwarf star it is today and become a red giant, its disc filling most of the earth's sky. An apocalyptic conflagration will destroy the surface of the earth and the seas will boil away. A billion years later the red giant sun will begin to shrink into a small white dwarf immensely dense and fuelled by its own gravitational collapse. After another fifty billion years its light will slowly fade and darkness will reinvade this part of interstellar space. In its final state the sun will have shrunk to the size of earth; a ball of crumbling iron, its usable energy all dissipated.

The sun gives life – and people – a chance. Its stable longevity is what has made the evolution of complex organisms possible. Many stars bigger than the sun burn themselves out long before life could get itself established on the surfaces of attendant planets. Other stars vary in their output of energy by hundreds of percent in periods of days or years while yet others are members of binary or group systems which would never offer the stable warm conditions of a regulated environment believed to be necessary for the successful evolution of life. Our own sun is a stable parent offering all that the earth needed to produce people and now gives us time to make a success or failure of our destiny.

We owe everything to the sun – the creation of our planet, the evolution of life and of ourselves, and now as a dependable source of energy that gives people time to look around and plan. We are products of and are intricately and essentially related to our environment. But our links with the universe do not stop at the sun, for we also have more distant ancestral roots in the stars. They are our chemical parents. Stars had to be born and die before we could be made. Our bodies of flesh and blood are built from molecules and the molecules from atoms such as carbon, nitrogen, sulphur, magnesium, iron, cobalt and many others. The origin of all these atoms illustrates once more how the emergence of human life is a direct result of the way the universe functions. The emergence of mind from matter is a product of the evolution of the universe itself.

In the earliest stages of creation just after the Big Bang there was only hydrogen and helium; none of the other ninety or so varieties of atom had yet appeared. All the more complex and essential atoms that make up our flesh and bones came into existence much later. Carbon, iron and the rest can only have been fused from simpler atoms in the nuclear furnaces that burn at the hearts of suns larger than our own sun. They would have been useless, however, if they had remained locked up in the infernal depths of giant stars. Fortunately they were liberated and shed through space in two ways. Firstly they were (and are) breathed out slowly from red supergiant stars such as Betelgeuse in Orion or Aldebaran in Taurus, a slow wind of particles fanned by the heat away from the glowing solar surfaces beneath which they are born. Others were shed through space more dramatically by cataclysmic star-bursts when giant suns became unstable and tore themselves apart in spectacular explosions called supernovas. One such event can be expected in a galaxy per century. The last supernova visible to the naked eye was in 1604, and the one before that in 1552 came to be known as Tycho Brahe's star after a famous astronomer of the day. It showed so brightly from the constellation of Cassiopeia that it could be seen in the blue sky of broad daylight. Perhaps the most famous supernova was the one observed and accurately recorded by Korean and Chinese astronomers in AD 1054. It too outshone all the other stars in the sky for several weeks and could be seen in the daytime. Today the disintegrated remains of the star have been called the Crab Nebula and the filaments of cloud composed of dust and gas are still expanding outwards from the initial explosion, at 1,100 kilometres per second. In 1967, Professor Anthony Hewish and Jocelyn Bell at Cambridge discovered a pulsar at the heart of the nebula, all that remains of the original star. It is a superdense object spinning at an incredible thirty revolutions per second and flashing radio waves in our direction.

The energy released in a supernova is so great that for several weeks the star shines with the brilliance of a billion suns or more. The heat generated is so great that heavy elements such as gold or uranium can be created by nuclear fusion. All the gold in every wedding ring on earth was manufactured in the fires of a supernova in the dim and distant past, for nowhere else in the universe is the temperature high enough for this cosmic alchemy.

The evolution of life could never have happened at the surface of the earth if it had not been for red giants and supernovas. Our own solar system, the sun, moon and planets, condensed through gravity from a cloud of dust and gas enriched by the solar winds of particles from red giants and by the bursting asunder of ancient massive stars at the end of their stable lives. In five thousand million years chemistry and biology produced order and pattern amongst the molecules and atoms lying around on the earth's surface. Plants, fishes, reptiles, mammals and finally men and women evolved from the rich supply of available material. The atoms that make up the skin and bone of the hands that hold this book and the eyes that read it are already billions of years old and have been through cosmic furnaces. Atoms do not grow old and die, not in the life span of the earth anyway. Nor do they spawn new ones. Recall that our bodies, bones, flesh and blood are built from what we eat. The iron in spinach was welded in nuclear furnaces by giant stars. But not only the iron; the carbon, nitrogen, sulphur and all the other atoms necessary for building organisms were disgorged from the depths of stars. What we eat comes from plants that grow in the earth or is the meat of animals that have consumed plants that grow in the earth. The plants themselves are created from minerals and other elements lying around in the soil. The soil contains the broken bits of rocks, the erosion of mountains and the dead remains of living organisms of previous ages. We are made completely from recycled bits, ancient atoms, which have been lying around or circulating through life forms for billions of years. They are the same old atoms that were there before the earth was formed and have been used again and again in mountains, rocks, plants and people. We borrow them briefly.

The Genesis story of creation calls the first man Adam, which means 'earth' because he was made from the earth of the ground. God made him from dust and at death the Prayer Book commits his body 'earth to earth, ashes to ashes, dust to dust'. What dust! The parable in Genesis is on the right lines but only scratches the surface of the mystery of reality. It took giant suns and supernovas to build the elements from which we are made. In the words of radio astronomer Professor Anthony Hewish 'without this vastness we would not be here'. The stars were once worshipped as gods but we now know that they are our chemical ancestors.

The great question facing biology today is 'How and where did life first appear?' We have already traced the source of the atoms needed to build living organisms to the hearts of giant stars. But how did they first come to unite in the complex molecular groups necessary for life? Was it by some remote accident or is the emergence of life in a chemical universe a normal and predictable outcome of basic scientific laws? There are no certain answers yet and the questions are heatedly debated. Christian theology has been tempted to cling to this enigma for the wrong reasons. It seemed that here was a phenomenon which science could not explain. Only a supernatural force could account for the appearance of life; it was God who bridged the gap between non-living molecular

groups and the first primitive self-replicating organisms.
Many Christians felt obliged to support what came to be
called the vitalistic theory since it safeguarded a role for God
in a universe which seemed progressively to be able to do
without God. Life was a mysterious extra ingredient im-
planted by God, the 'life-force' of Shaw or the 'élan vital' of
Bergson. But vitalism and theologies of supernatural intervention are now held to be a weak support for faith. God is
better found through what we discover about the world and
not in what we find difficult to explain. The God of the gaps is
a dead idea.

The true glory of matter is that vitalism is an unnecessary
hypothesis. The parable of life as a breath breathed by God
into creatures of dust was appropriate in a prescientific age.
So long as we remember that it *is* a parable and uses the
language of poetry rather than science then it can still make its
deeper point that all things are dependent on God. But the
great wonder revealed by science is that it is the laws of nature
which turn mud into flowers. Life emerges in the universe
because that is the way the universe is. The human spirit is the
product of chemical and biological evolution and, at root, the
laws of physics. The human mind is the most remarkable
blossom on the tree of life, for in us chemistry opens its
petals of wakefulness. Was the origin of life 'in some warm
little pond' as suggested by Charles Darwin? Experiments
with prebiotic 'soups' can still only produce tentative recon-
structions of the first stages of the evolutionary process that
gave rise in the first instance to self-replicating organisms.
Mixtures of methane, ammonia, water and hydrogen molecules energized by an electric current the equivalent of a
lightning bolt in the atmosphere of the early earth, or bom-
barded with ultra-violet light which in the first era of our
world would have been able to penetrate the primitive atmosphere, can produce amino acids, the building bricks of pro-
teins and thus of life. The origins of the genetic code by which
each generation passes on to the succeeding one a plan of how
to build the molecular architecture of an organism is, however, another matter and remains baffling. The simplest living

creature is so complex that it could not have been constructed by accident in one go. Fred Hoyle in an entertaining image has suggested that the odds against life emerging this way are equivalent to the odds against a tornado assembling a jumbo jet by blowing through a scrap yard. He estimated the odds to be 10^{40} to one. Hoyle compared the claim that the leap from chemistry to life was accomplished by a natural process on the surface of earth to Aristotle's belief that fireflies were born from morning dew and small animals from the mud.

Fred Hoyle (with his colleague Chandra Wickramasinghe) has his own controversial theory about the origin of life from space.[5] Unfortunately, owing to the fact that Hoyle tends to produce outlandish theories which many scientists see as merging with his science fiction writing and also because he is not a biologist, his scenario has been received badly in some scientific quarters. The theory itself is plausible and has entertaining possibilities. Astrophysical research has revealed spectroscopic evidence of many of the complex molecules needed to build amino acids spread thinly throughout the clouds of gas that drift between the stars. Hoyle's interpretation of some of the evidence (to the extent, for example, of detecting the absorption spectra of the bacteria *escherichia coli* in space) is considered to be both extravagant and unwarranted by many other astronomers. Nevertheless the basic materials for life are already there, spread out through the universe. Hoyle speculates that on the surface of sticky dust grains or in the icy heads of comets these molecules may begin the slow evolution towards primitive life forms. When comets collide with planets they discharge their living cargo and the seeds are sown. Later comets continued to enrich the environment with viruses, and Hoyle even speculates that 'flu epidemics and the common cold are visitants from space. (Not a new idea – 'influenza' being the Italian name for those epidemics of fever thought to be the effect of stellar 'influences'.) Who knows what might be lurking in the tail of Halley's comet as it swept round the sun in 1986!

We might argue that whereas the warm, regulated conditions on the young earth offered a credible environment for the emergence of life, the almost absolute zero conditions of empty space are surely overwhelmingly hostile. Yet life on earth provides its own examples of organisms that can survive in the most rigorous and adverse conditions. Not far from the Galapagos Islands in the Pacific there are bacteria which happily live their lives two miles down in the ocean, well beyond the reach of sunlight, in water that reaches temperatures of four hundred degrees centigrade where it emerges from hydrothermal vents in the sea bed. At the other extreme there are organisms which survive in the permanent sub-zero temperatures of the Antarctic. Hoyle claims further evidence for his theory, that life originated in space and not on earth, from meteorites. One interpretation of some of the minute features found in stony meteorites called carbonaceous chondrites is that they are fossilised micro-organisms. If this view is correct then these primitive life forms were fossilised long before they fell to earth from the sky.

The reservations many scientists have about some of Hoyle's speculations should not however be allowed to obscure the main thesis of his book *Lifecloud*. The major building blocks of life, even perhaps the simplest life forms, may well have originated in space and rained down on the planet as alien organic pollutants. But Hoyle has only back-dated the problem. Life may be spreading through the universe as prolifically as weeds in a garden but we still do not know how the leap was made in the first place from dead chemistry to living organisms. Nor do we know if it happened only once and spread out from that seed event or if it occurs throughout the universe in innumerable sites.

Many biologists are confident that even at the preliving molecular level a process of natural selection took place. The distinction between living and preliving may be purely semantic at this point but it seems a perfectly rational conjecture that the first single celled creatures, our remote ancestors, were themselves the climax of millions of years of haphazard evolution amongst molecules. This confidence that one day science will be able to offer a detailed explanation of the mechanics of the process is something that Christian theology ought to be able to share. An outmoded theology which smuggles God into the system to interfere with the laws of nature is much less sophisticated than a scientific account which sees all the emergent richness of life as the true glory of chemistry. The scientific account has the touch of *true* mystery.

The more we look at the universe the clearer it becomes that all things are interconnected. We are a consequence of the cosmos and cannot be considered in isolation from it. This new insight is a contribution science makes to our search for meaning. Julian Huxley, an atheist by confession, linked humans to the evolutionary process with a memorable phrase. In the introduction to *The Phenomenon of Man* by a Jesuit theologian and palaeontologist, Teilhard de Chardin, he wrote 'in [the human] evolution became aware of itself.' In our limited experience of the universe it is only in the complex arrangements of atoms and molecules that form the nervous system and brain of a human being that self-awareness has emerged. We now appreciate that these atoms and molecules have a history vastly more ancient than the period in which life has appeared on earth. Their creation was an essential preliminary to evolution. Julian Huxley's phrase can be taken back a stage further. In the human, not only evolution, but chemistry becomes aware of itself. Atoms wake up to self-knowledge when they are organised as people.

In the first eras of the scientific quest to comprehend the heavens we suffered from what has become a familiar problem – the misreading of statistics. Insecurity entered human consciousness as it appeared that we are no more than insignificant specks of dust in some forgotten corner of an alien universe. Cosmic agoraphobia afflicted our souls. A new complexion is put on the matter, however, when we comprehend the ancient pedigree of our chemistry. People are not to be counted and assessed in terms of an odd peripheral biological statistic. In people, matter realises its true potential. The description of an atom must include an account of what emerges when it combines in an organised way with other atoms. It is no longer unreasonable for us to begin to suspect that we are what the universe is for. We are the eyes of the universe; the most awake bit of the cosmos known to us. The adjustment we need to make to a new scale of things has a further consequence. In terms of our chemical and evolutionary ancestry humankind is incredibly new. Religion must adapt to this and help us to look forward to what we may become. Christ, Buddha, Lao-tzu, Confucius, Moses are our contemporaries not misty figures in the past. We may be tempted to think that their days are dead and gone, that we live in a post-Christian and sometimes godless world. But our view of time has been hindered by a false scale of history. The human venture, after billions of years of preparation, has only just begun and *now* is the time that we need spiritual guidance. For a Christian the revelation of God in Christ is for today and tomorrow, not a fossilised idea from yesterday.

How new we are is easily illustrated. There is a scrubby looking creosote bush in the Arizona desert which has been alive for 11,000 years.[6] The same plant, slowly spreading outwards century by century, has survived through the whole period of human history from the building of the first village cities at the end of the last ice age up to the flush of technology that put a man on the moon. The rise and fall of all the great empires on earth and the flowering of civilisation is spanned by the life of one slow-growing shrub. Other examples can reinforce the discovery of just how recently humankind appeared on the planet. Sharks have been hunting the oceans of the world for 150 million years, while the human race has been around for less than two percent of that time. Cockroaches were scrambling happily across the Weald in southern England 100 million years before there were any kitchens for them to infest or cookers to hide behind.

The fossil-collecting child on a Dorset beach examining with wonder the spiral pattern of a 150 million year old ammonite in a pebble is a symbol of where humanity stands now. We are newly awakened creatures, only just ascended from the animal kingdom. With the insights of the twentieth century we are only just beginning to grasp who we are and how we got here. The astronomer gazing at the spiral galaxies of stars whose light has taken hundreds or even billions of years to reach his telescope may experience the same wonder as the child contemplating the fossilised nautilus shell. Both the fossils on earth and the stars in the sky have important things to tell us about ourselves.

'The heavens declare the glory of God' wrote the psalmist 'and the firmament showeth his handywork.' The anonymous poet who wrote Psalm 19 lived, we presume, amongst the hills of Judaea two and a half thousand years ago, and the sky which filled him with wonder is exactly the same sky which we see today. The constellations the psalmist saw are the same we gaze at in the twentieth century. Optical telescopes and radio antennae, satellites and space probes are giving us a detailed description of the heavens. We can now piece together an account of their evolution, age and dimensions, and yet there is not one iota of information or cosmological model which need call in doubt faith's conviction that they are the handiwork of a divine creator. The heavens have played an essential part, too, in the emergence of life and mind on this planet. Through the atoms they have fashioned they have found a voice to praise God. For some people today the heavens *still* declare the glory of God. Even more so.

3. GOD AND COSMOLOGY

The Christian faith today has to be understood and practised within a new cosmic framework. The small stage offered by the Bible, which once was comforting, is no longer adequate and is being replaced by something more realistic. According to early Christians no more than seventy-seven generations of people separated the birth of Jesus in Bethlehem from the creation of Adam the first man in the garden of Eden (Lk 3:23–38). In their view the play of history was almost done and with the appearance of the Messiah in Israel the curtains had been raised on the last act. Biblical history, as far as they were concerned, was world history. The earth was a small place protected from the waters of chaos that lay all around by a tent-like roof, the sky, and the sun and moon were added for convenience to give illumination. The stars were nothing, merely an afterthought, tiny pinpricks of light. Adam and Eve tempted by the snake spoilt the paradise on earth which God had created, planted, watered and peopled in six days. They were disobedient and fell from grace. A few generations later God sent a flash flood, destroyed almost everything and made a second start with Noah and his family. Again things went wrong and God sought to put them right through the mediation of the children of Israel, a chosen race. From this race a messiah emerged, he lived and died and rose from the dead and then promised his followers that he would return. The first generation of Christians lived in daily expectation of the second coming, the parousia, when Christ would reappear and God would create a new earth. The end of the world for these Christians was close.

This account of 'salvation history' served Christian society well for almost two thousand years despite the delay in the parousia. In the seventeenth century Archbishop Ussher gave it quasi-scientific support by calculating that the world began in the year 4004 BC precisely, and orthodox Jews still reckon their calendar from the first day of creation by a slightly different method so that 1986 is the year 5747. This short framework of history continues to have a strong grip on some minds in the late twentieth century. A vociferous body of opinion within the Christian church still clings to it tenaciously as the only correct world view. But modern astrophysics confidently tells us that the world is five billion years old not just six thousand and that the universe is three times more ancient. It began in the remote past with a big bang and a fireball of light so hot that for almost a million years even the simplest of atoms could not exist. The mould has been broken. Modern cosmology has shattered the biblical time scale for ever.

One truly remarkable thing about our knowledge of the universe is that we never see it as it is but only as it was. Astronomers are forced to study the past. This makes little difference in the case of the moon for the sunlight bouncing off its rocky surface reaches us just one and a quarter seconds later. Even events on the sun are seen from earth with a delay of only eight and a half minutes. But what we know of anything beyond our solar system is always well out of date. Information via the various parts of the electromagnetic spectrum, light waves, radio waves and so forth, take four and a half years to arrive from the nearest star. Every constellation we see in the night sky is a sort of time tunnel, since each star is at a different depth in space and the light reaching our eyes in fact comes simultaneously from different eras of the past. We see one star as it was in the eighteenth century, for instance, while its neighbour shines down on us from the sixth century.

The lovely star clouds visible on summer evenings in the broad ribbon of the Milky Way where it runs through the constellation of Sagittarius shine faintly from several thousands years ago. They are the light of millions of suns clustered together in the direction of the centre of our galaxy. Anyone on a Mediterranean holiday in August is particularly well placed to see this beautiful patch of sky. But the naked eye can see even further back in time than that. The most remarkable filament of the past visible without optical aid is best seen in late summer or autumn. The four stars of the square of Pegasus begin to dominate the southern sky in late evening. To the left of them lies the constellation of Andromeda. The stars in Andromeda are all a hundred or so light years away and therefore are shining from the relatively recent past. Embedded in Andromeda however there is a ghost-like blur of light, no more than a tiny smudge of a glow. It is popularly known as the Nebula in Andromeda but on star maps simply as M31. Earlier this century the true nature of the nebula was discovered. It is a vast star system in its own right, lying far beyond the boundary of our own galaxy. An amateur astronomer without the aid of a telescope or even binoculars (though they help) can contemplate that great swarm of a hundred billion suns as they were two and a quarter million years ago; it has taken the light so long to reach us. Humankind had only just emerged from other primates when the light arriving today set out on its journey.

As we look out from earth into the universe we gaze back through the layers, as it were, of a chronological onion. The greater the distance of an object the further back in the past we see it. With the most powerful telescopes astronomers are able to see galaxies as they were ten billion years ago when the universe was only a third of its present age and long before the formation of our solar system. Telescopes soon to be put into earth orbit above the obscuring effects of the atmosphere will probe even further into the past, almost to the beginning of the universe and of time. The beginning itself will always remain hidden from us, however, since it lies beyond the horizon of what is observable.

The insights of science, particularly of astronomy, require that the Christian today develop a new creation story and a fresh understanding of the nature of God. He will also need a new concept of the inspiration of scripture. In what way is the

Bible true and at what level is the truth to be found? Already the six day creation story and the garden of Eden parable are being read with new eyes. Theologians no longer see any sense in speculating about the geographical location of Eden or attempt to dovetail the episodes of creation in Genesis with evolutionary theory. The stories in the opening chapters of the Bible are vehicles for creedal statements about the meaning and purpose of humanity and the universe. They are not history or science and indeed how could they be? Scientific method involves a combination of theory and experiment, of calculating and then testing the consequences of hunches. Such an empirical probing approach to reality was undeveloped when the Bible was written. The scientific mind had not yet awakened.

Most people are very hazy about the origin and authorship of the Bible. If we are to develop with any confidence a new framework for the Christian faith then it is important to understand first of all how the old framework came about. Then we may find where the truth lies in scripture and where it lies today in terms of modern cosmology.

The majority of theologians explore the authorship and origins of the book of Genesis on the assumption that it contains myth. In this context 'myth' is a technical term and it is essential to know how it is being used. The Greek root of the word means no more than an 'utterance' or later, by implication, a 'story'. Through time 'myth' has spawned a number of meanings two of which stand in opposition to one another. In everyday language a myth has become an untrue story or a misconception, a false tale to be rectified or slanderous accusation exposed. In theological discourse the meaning is quite different. A myth is a story with a deep significance, embodying a truth, and it may take the form of an historical tale. The account of God creating in six days, for example, is a myth. So, some would argue, is the exodus of the Israelite slaves from Egypt. In this latter case the myth has at its core a thread of history. Some Israelites, maybe only a few thousand and perhaps only six tribes, escaped from Egypt against all odds. Inspired leadership and a sense of destiny brought them to a land where they could settle. Many of the details of the escape and the wanderings may have been invented, exaggerated or drawn in from other stories. Some of the miraculous happenings may have been overstated with hindsight. But none of that really matters for the great power of the story is in its conviction that God saves his people and will always lead them from slavery to freedom. In Genesis, however, the mythical form of the literature is more obvious. The details of the six day creation need not be taken literally because the meaning of the story is more profound. God is creator and God has a purpose for the world and all people. For a theologian, a myth is a vehicle for truth, not falsehood, but the truth may lie at a different level than the superficial details. Myths of course are not unique in this respect but have affinities with other didactic or literary devices such as parables, analogies, metaphors and symbols. In the case of myth however the subject matter usually has to do with some basic religious truth.

The Genesis stories as we now have them were put together, it is generally assumed, in the sixth century BC by refugees from Judaea exiled in Babylon. Textual criticism reveals that they are a beautifully edited patchwork of materials from a number of different sources. The name of God, for instance, varies from Elohim to Yahweh depending on the origin of the text. Behind this variety of sources lies a long history of oral tradition about which we can speculate only tentatively. The sixth century BC was both a low and a high point in Jewish history. The Babylonians had destroyed the last remnants of Judah and laid waste to the city of Jerusalem. All the Jews' hopes for the future had come to nothing. God who chose them, promising that he would make them a great nation, now seemed to have deserted his people. Even worse, the prophets said they were being punished. The Temple, the centre of their cult and their faith had been torn down and all the leaders of society carried off into exile to live by the river Euphrates. 'By the waters of Babylon there we sat down and wept' wrote a contemporary psalmist (Ps. 137:1, RSV). But punishing events sometimes make people think and take stock. Out of the ashes of this national holocaust rose the phoenix of scripture.

One of the finest of the exilic writers was an anonymous follower of the prophet Isaiah who had lived two hundred years earlier. He has come to be known as Second Isaiah. Sublime passages from his pen are incorporated in the book named after his predecessor. A truth that had been known to the children of Israel struck him with fresh force. Only God is God. Second Isaiah is uncompromising in his monotheism. He satirises and mocks the stupidity of the craftsman who cuts down a tree, uses some of the wood for a fire to keep him warm and cook his food while from the rest he carves a god which he covers in gold and worships (Is 44:13–20). He laughs at Bel and Nebo the splendid carnival gods of the Babylonians who, gilded and painted, had to be carried through town by beasts of burden (Is 46:1). Yahweh, by contrast, carried the children of Israel throughout their history:

> Who was it measured the water of the sea in the hollow
>   of his hand
> and calculated the dimensions of the heavens,
> gauged the whole earth to the bushel,
> weighed the mountains in scales,
> the hills in a balance?

> Who could have advised the spirit of Yahweh,
> what counsellor could have instructed him?
> Whom has he consulted to enlighten him,
> and to learn the path of justice
> and discover the most skilful ways?

> See, the nations are like a drop on the pail's rim,
> they count as a grain of dust on the scales.
> See, the islands weigh no more than fine powder.
>   (Is 40:12–15, JB)

While Second Isaiah was writing his satire of contemporary religious practices and realising the cosmic implications of his monotheism, others were editing the various strands of their

religious tradition. Much of the Old Testament as we now
know it comes from this period of history, in particular the
first five books of the Bible known as the Pentateuch or
traditionally, though incorrectly, ascribed to Moses. The
creation myths of Genesis are an important and significant
part of this collection. It was in the raw atmosphere of exile
when circumstances had brought a peculiarly self-conscious
people to their knees that new spiritual insights burst upon the
Judaean refugees. A new spiritual awareness focused their
attention on the two great mysteries, God and the individual
(Jer 31:31; Ezek 18). Human consciousness made one of
those rapid leaps forward that seem to punctuate the history
of ideas.

The editors of the Pentateuch were totally committed, like
Second Isaiah their contemporary, to the belief that the
beginning and the end of history lay with God. God, who had
chosen their race and laid on them the task of awakening the
world to divine being and nature, was the God who created
everything. It was logical, then, that they should start their
history with the words 'In the beginning God created the
heavens and the earth . . .' (Gen 1:1, JB). The story that
follows, a myth in theologians' language, elaborates the
editor's fundamental conviction that God is Lord of the whole
earth and the heavens and that humans are the most valued
creation. The day to day details are no more literally true than
that God actually did measure 'the water of the sea in the
hollow of his hand' in the metaphor of Second Isaiah. It is also
very doubtful if they invented the story from scratch (or that
God dictated it from on high). Most probably they retold a
Babylonian creation story with which their earliest readers
would have been familiar. The editors brought a number of
Jewish features to the retelling of the myth. The sun and
moon, the gods Shamesh and Nana to the Babylonians, are
demythologised and reduced to the status of lights in the sky.
Internal evidence suggests that the original story was eight
verses for eight days. Days five and six seem to contain a
double dose of creative activity. This conflation left the
seventh day free for God to rest – powerful reinforcement for

what had become by then the regular Jewish observance of the sabbath. During the exile, the greatest threat to the peculiar Jewish religious culture was that of assimilation. Kosher food laws, circumcision, synagogue worship and the sabbath rest were essential if the exiled community was to keep its special identity. A six day creation myth which culminates with God resting on the seventh day had its own propaganda overtones for the faithful.

The story of Adam and Eve and the garden of Eden is part of a second creation myth which follows on from Genesis 2:4. It is from a separate source and originally was unconnected with the first creation story. The focus here is upon human disobedience, pain and suffering, a mythological answer to those who first began to question why a good God could be responsible for a world so ill at ease. The mythical tree, a tree from a fable rather than a forest, bore the mysterious fruit of the knowledge of good and evil. (Eating from it, it could be argued, was part of the awakening God intended for humanity.) Woman, made from Adam's rib — a male chauvinist element here not present in the six day creation story — has a role to play in the temptation and the fall. The myth of the paradisiacal garden offers a vision of what life could be like and makes a comment on what we do with our freedom. The description of Eden may have been inspired by some beautiful oasis in the desert but the garden was no more a real place in a geographical or historical sense than it is a scientific fact that God put the sun and the moon in the sky on the fourth day after creating plants and trees on the third day.

The literal interpretation of Genesis has a relatively recent history. It coincides paradoxically with the rise of scientific method and will one day be reviewed with hindsight as a temporary aberration in theology. Exactness in science led to a craving for the wrong sort of certainty in matters of faith. The Reformation caused the Christian world to focus its attention on scripture, as protestant movements wishing to rid the church of medieval accretions and abuses turned to the Bible as the true touchstone of faith. In 1545 the Roman Catholic tradition, pricked and quickened by this protest of

Christian conscience, convened the Council of Trent. This council, not wishing to be outdone by reformers, declared God to be the author of the Old and New Testaments. Such a statement is of course open to a number of interpretations depending on your view of divine inspiration. One of these leads to the unfortunate fundamentalist stance. If God is author then it could be argued that the writers of the Bible are no more than mindless secretaries taking down what is dictated. God said the world took six days to create and so any evidence to the contrary from biology that evolution has happened or from cosmology that the universe is fifteen billion years old is wrong by implication. The major and most exciting insights of modern science are rejected as godless frauds.

The majority of theologians favour a different interpretation of divine inspiration. Over four hundred years after the Council of Trent, the Second Vatican Council, which began in 1962, clarified the Roman Catholic point of view with the statement that God *inspired* the authors of the Old and New Testaments. God's is the inspiration but the words are theirs. Just as a poet will seek the right words to convey an experience, so the numerous authors of the Bible expressed in their own way, and in the thought forms of their time – using the stories and historical material at hand — the truths they believed had been revealed to them. A Babylonian creation story was pressed into theological service, moulded by its new redactors, or editors, and used to proclaim their dawning vision of uncompromising monotheism. Like the creator in whose image they believed themselves to have been made, they used their material creatively to give substance to their statements of faith. In the same way today's theologians will turn to the scientific models of the origin and evolution of the universe. Astrophysics and theology each offer insights into reality and it ought not to be surprising if, through dialogue, each tradition is found to enrich the other.

It is often forgotten that inspiration is not the monopoly of the writer. God's Spirit is as much with later generations who read, mark, learn and inwardly digest as it was with those who wrote and edited the Bible. We who live in a scientific era may be guided by God in our reading of the Genesis creation story and perceive new insights not available to the early writers. Like Newton we have the advantage of standing on the shoulders of giants, spiritual colossi.

Modern astrophysics gives us a new context within which to comprehend our life, a new framework for our faith. People have always sought some sort of an explanation of how we came to be here. The Smith River Indians of California imagined the world to have been scratched up from nowhere by the legendary dog coyote; Japanese storytellers had the notion that the gods stood on the bridge of heaven and stirred the ocean beneath with long reeds and thus created solid land; the Aryan invaders of India taught that Brahma made the world, while for the Greek atheist philosopher Democritus everything was the result of atoms exploring aimlessly all possible configurations through an infinity of time. Imagination, guesswork, untutored speculation all played a part in these stories and explanations. In the twentieth century for the first time humankind has within its grasp the basis of a correct answer. This answer may seem less personal and more remote at first than the older mythologies but as a story which explains how we came to be here on Planet Earth it fulfils some of the functions of earlier creation stories. Some writers even go as far as to call cosmology the modern myth. Both practical astronomy and mathematical and astrophysical theory point to one model of the evolving cosmos and its origins. With a poetic echo of the Genesis myth, the universe began in a fireball of light, the Big Bang.

The dying reverberations of that ancestral explosion can still be heard in the form of radio interference once suspected to be merely the effect of pigeon droppings in a Bell Laboratories radio receiver. $3°K$ microwave radiation, predicted by theory and measured in practice, floods space with a gentle whisper from every direction. More evidence for this latest scientific model of the universe is provided by the relative motions of all but the most local galaxies. Throughout space, galaxies of stars are flying away from one another as the universe continues to expand fifteen billion years after the primordial violent event set it in motion in the beginning. The further we probe into space with optical or radio telescopes the more the electromagnetic radiation is shifted towards the red end of the spectrum indicating that the source of light or radio waves is receding from us. This 'red shift' is such a ubiquitous phenomenon that it has come to be used, with a few exceptions, as a reliable method for estimating cosmic distances. Some galaxies are so remote that they are retreating from us at many tens of thousands of kilometres per second.

No precise figure can yet be given for the age of the universe, though an upper limit of twenty billion years is generally accepted as being within the right order. Calculations are based upon estimates which vary, such as the Hubble constant which puts a figure on the rate of expansion of the universe. Amusing news headlines in 1980 suggested that scientists had knocked billions of years off the universe's age when they readjusted the Hubble constant. This immediately raised problems because on other grounds globular star clusters, among the most ancient denizens of the galaxy, were reliably being dated with ages that made them several aeons older than the universe itself. Such temporary inconsistencies are bound to be ironed out as theory and observation build up a more accurate picture. There are far fewer doubts, however, about the first few minutes and cosmologists discuss the first few microseconds of the Big Bang with amazing confidence. Nobel prizewinner, Steven Weinberg, has written a whole book dedicated to 'the first three minutes'.[1] Even this level of microchronology has its eras and epochs and the most recent speculation is about the period that lies before the first 10^{-35} of a second.

The whole universe grew in a flash from a 'seed' known, to mathematicians of the process, as a singularity. Not only light and heat, and later matter, emerged from this seed but also space and time. Space and time are inextricably linked to each other and neither can be separated from the universe since they are aspects *of* the universe. Before the Big Bang there was no 'where' or 'when' and there are obvious philosophical problems in talking about the 'beginning of time'. To ask 'when did time begin?' involves a nonsense. The significance of that brief era in the beginning (less than a billionth of a blink of an eye) is in the manner of the universe's expansion from the singularity. It is here that mathematicians account for the smooth uniformity that astronomers find throughout the cosmos as they view it today. Whichever way they look out into deep space the distribution of galaxies and the intensity of the microwave background appears the same with a deviation of less than one percent. In that 'inflationary era', as the epoch has been called, all the forces of nature, (strong, weak, electromagnetic and gravity) may have been one single force. By 10^{-35} of a second the universe had expanded and cooled sufficiently for the forces to separate out as distinct independent phenomena. The temperature of the expanding fireball was still trillions of trillions of degrees Kelvin and only after a million years did this exploding cosmic furnace become cool enough for the first atoms of hydrogen and helium to form from more fundamental particles such as protons, neutrons and electrons.

When the Big Bang theory was first being considered it had a competitor. The alternative view, called the Steady State or Continuous Creation theory, proposed that the universe had no beginning, is eternal and continually replenishes itself with new matter which emerges between the galaxies as they drift apart in space. Superficially this eternal process of continuous creation seemed to dispense with the idea of a creator God (an illogical conclusion because there is nothing inconsistent with the concept of a God creating eternally). Many theologians consequently espoused the Big Bang hypothesis in that it fitted neatly with the opening words of Genesis 'In the beginning'.

In the scientific community the Big Bang model of the cosmos has almost universal support as the best hypothesis. There is much less certainty however about where the universe goes from here. Judgement day, the day of reckoning which would see the return of Christ and an end to world history, is not just round the corner as traditionally expected by Christians. Although some sects continue to announce that 'next year' the world will end, the majority of Christians long since ceased to expect judgement in those terms. We may face extinction at our own hands, but that was not quite what the prophets of the Old Testament meant when they warned that the day of the Lord was close. Modern cosmology which pushed the 'beginning of time' back a millionfold has done the same for the future. There will be no neat end to cosmic history in our lifetime or even within the lifetime of the human race.

Two possibilities face the universe in the distant future but scientists do not yet have enough information to know which is most likely. What eventually happens depends upon how much material there is in the universe. If there is sufficient then the combined gravitational attraction of all this matter will slow down the rate of expansion until it finally stops. On the principle that 'what goes up must come down', everything will begin falling together again. In some remote era hence, many times older than the current age of the universe, there will come the Big Crunch. Collapsing back into a superdense state the whole of the cosmos will 'vanish', space and time once more warped into a singularity. Nothing will escape the doom of this black hole, the ultimate cosmic coffin. Mathematicians can calculate the critical mass needed to halt the outward expansion from the Big Bang. So far astronomers have located less than twenty percent of this mass by adding up all the material that makes up visible stars in galaxies, all the estimated dead stars, clouds of interstellar gas and dust and possible quantities of massive neutrinos. Even so, many are hopeful that one day the critical mass will be accounted for and then we will know for sure that in perhaps 100 billion years the universe will finally collapse back on itself.

The alternative scenario is that the universe will go on expanding for ever, slowly using up all its available energy, until it is spread out into the desolate silence of an exhausted void. There will be no final moment, no cosmic cataclysm to mark the end, just a timeless freeze. Darkness will reign.

Big crunch or slow decay, either way the death of the universe is inevitable. Everything in this universe eventually wears out, even the cosmos itself. We know from experience how true this is for washing machines or cars, shoes or people. The dying Buddha underlined it in one of the last things he said to his grieving disciple Ananda: 'Is it not true Ananda that in this world all things that are put together will one day fall apart?' The comment applies as much to the composition of the atom or to the dynamic structure of a star as it does to our frail bodies. In the long run death rules everything. Many people would argue that the death of the universe is guaranteed by the second law of thermodynamics sometimes called the law of entropy. It is one of the most comprehensive generalisations in the whole of science and tells us that the total entropy, or disorder, of the universe increases with time. The law is of such fundamental importance to our understanding of the world that C. P. Snow held it up as a general test of literacy. Not knowing the second law is a bit like not having read any Shakespeare.

With the Big Bang a vast amount of energy was wound up into the cosmos. Relentlessly it is being used up and degraded, since all change, whether it be the burning of a star or the growth of a tree, contributes to energy's decay. Just as desk tops and children's bedrooms are prone to become disordered unless energy is put into tidying them up so there is an irreversible trend in the universe itself towards chaos. In the Big Bang the universe was given order but it is ceaselessly unwinding and using up the high quality energy made available. According to the principle of the conservation of energy enshrined in the first law of thermodynamics, the total energy of the universe remains the same. The problem highlighted by the second law is that this energy becomes increasingly useless. The solar energy locked in coal for instance is dissipated in heat, light and noise as it burns. A film of the burning fire could be played backwards but not the physical process in the real world. Ultimately all change in the universe runs in one direction towards chaos and useless energy. The decay is irreversible. In a remote epoch from now, long after the lifetime and death of our sun, the universe will die, crushed beyond anything we know in a black hole or drifting towards total stagnation and freezing in darkness.

Between the prologue and the epilogue of this vast cosmology humankind has its fleeting life. In such a transitory world the human spirit seems even more ephemeral. The great question for the Christian faith in this age concerns the *status* of that spirit. Is it a passing phenomenon, a mere temporary bloom as short lived as the morning glory flower? Or does it qualify for some new status for which such words as immortality and eternity are employed?

The creation myth in Genesis culminates on the sixth day in the creation of human beings made in the image of God:

> And God said, Let us make man in our image, after our likeness: and let them have dominion over the fish of the sea, and over the fowl of the air, and over the cattle, and over all the earth, and over every creeping thing that creepeth upon the earth.
>
> So God created man in his own image, in the image of God created he him; male and female created he them. (Genesis 1:26–27, AV)

Modern cosmology gives a new order and pattern to creation. Where do men and women stand in this new framework and what sort of value do we put upon the human spirit in our theology? Is it possible that the appearance of life and mind on the surface of this planet indicates that a victory has been won over that dragon of chaos, the law of entropy?

It has been argued by Arthur Koestler and others that the evolution of life on earth, and subsequently of humankind, reverses the direction of entropy.[2] The human brain is the most complex physical structure known to us in the universe. The increasingly complex chemistry of living organisms brings order out of a chaos of molecules and the human brain is its most recent and sophisticated product. Life is a rising tide running victoriously against the ebb flow of entropy. A new word 'Negentropy', embodying a double negative, was coined to indicate the positive creative dynamic of the process of evolution. Unfortunately, stated this way this optimistic view of universal change is flawed. No physical process however complex can ever stem the inevitable flow of corruption and decay. The evolution of life brings some order into the material world but it amounts to no more than a temporary eddy in the general trend towards disorder. The law of entropy only applies to closed systems which have no new energy added from outside. Domestic fires will only continue to burn if more coal is added. Likewise life can continue to change and evolve only so long as it draws on new supplies of energy from its environment, the sunlight, air and soil. The machinery of evolution is driven by the sun. So long as the sun contributes new energy each day the process of evolution will continue to flourish unless some disaster blows it out. Life on earth does not evolve as a closed system and so for the time being the laws of entropy seem to go into reverse, as the chemistry of organic matter becomes increasingly complex. In fact, however, just by living and eating we are contributing all the time to the decay of energy. There is no way out by this route and in the end the dragon of chaos will triumph.

Many people will feel themselves constrained, reluctantly, to conclude that the spiritual side of humans is totally and always dependent on a physical brain and nervous system. The laws of nature working on the materials created in stars with the opportunities offered by chance, given time, generate the human soul. But the human spirit, or soul, remains in this view simply a fleeting epiphenomenon of the physical world, a mere temporary bloom. If so, then the universe for all its groaning (to use the pauline terminology of the letter to the Romans) will do no more than give birth to temporary offspring. The law of entropy will ensure that at the end of time nothing of spirit will remain and God, if there is a God, will be left with only a memory.

Against this courageous but ultimately pessimistic view the Christian gospel speaks of eternal life and of 'the glory, as yet unrevealed, which is waiting for us'. (Rom 8:18) It is a gospel of hope even in the face of decay, corruption and death. It sees the human spirit as a new creation emerging from the material matrix of chemistry and biology but capable nevertheless of independence. The image of God in humanity is a potential condition waiting to be realised. We are more than the sum of our chemical parts and our spirit is a new thing with its own unique reality before which the vocabulary of science falls short. The human spirit, an emancipated product of matter, has power over matter.

The Christian faith cannot be accused of making a desperate last ditch stand against the inroads of unbelief when it finds mystery in humanity. Far from grasping at a gap in our understanding of the natural world — a final opportunity to find a slot for God — it is, rather, responding to a genuine experience of wonder. When our physical machinery and the electrochemical activity of our brains are methodically described and explained we are still left with mystery. The mystery is not beyond the comprehension of humanity but it requires a different language to describe it than the language of science. Poetry and theology, analogy and symbol are better suited to the task. The biologist or behaviourist can give a scientific account of a human being but cannot *as* a scientist describe the potential value, beauty and creativity of the human spirit. But then a scientist is not just a scientist, he or she is primarily a person, a unique mystery in his or her own right.

Science has its own comment to make on the status and significance of humanity. Four hundred years of scientific enquiry have methodically dispelled mystery after mystery, leaving very few dark corners unexplored. In the near future we may know everything there is to know about physical reality. In one of the boldest footnotes of twentieth-century science Peter Atkins made this optimistic prediction:

> Fundamental science may be almost at an end, and might be completed within a generation. Such views have been held before, but people mistook smallness for simplicity.

Only when structure has been peeled back to the point where it ceases to need further structure – when it has been peeled back to the point of extreme simplicity, such as lack of spatial extent – can we be confident that we have reached the end . . .[3]

Yet a clear scientific explanation of the universe does not cover all that has to be said: with knowledge comes wonder. Mystery is being rediscovered through the revelations of science, and faith begins to find support from what we know rather than from our ignorance. A scientific account of the universe can no longer treat people as though they were a peripheral oddity, a minor entry in a cosmic encyclopaedia of knowledge. The vastness and antiquity of the universe tended in the early days of scientific enquiry to reduce people to nothing. Today the situation is being reversed and one sign of this is the formulation of a principle known as the anthropic principle.

The anthropic principle (from the Greek 'anthropos' for human) reveals how intimately we are connected to the universe we inhabit. We have already noted that the atoms which make up human bodies were welded in the nuclear furnaces of giant suns. Inevitably their liberation from those infernos and subsequent collation and evolution on a warm stable planet took billions of years. The anthropic principle, first articulated by the astrophysicist Brandon Carter[4] in 1974, goes further than observing that people are made from 'star dust' and argues that given the way things are humans were bound to appear. Our understanding of physics and chemistry, of the way stars 'burn' or of the way fundamental particles relate to one another in atoms all point to one conclusion. This universe was designed to make people. We are not the result of a bizarre and unpredictable accident in an inhuman cosmos; we were bound to appear on the scene just as surely as hydrogen and oxygen make water or that water flows downhill. A description of the cosmos is incomplete without an account of the human because this is the inevitable product of matter in a universe such as this.

The universe was made to make humans. This conclusion is more of a hunch than a proof that life has a purpose or that God created everything. We only know of one universe and many people would think it improper to argue that its pattern implies purpose or intention. The universe *has* produced people and so clearly its conditions must be just right *for* producing people. Nevertheless, though dubious statistically, the hunch has a certain persuasive power. No observers would ever have appeared on the cosmic scene if any one of a great many precise conditions had not been satisfied. It is as though someone had finely tuned the values of the fundamental forces that bind the atom or generate solar systems with humans in mind. They give rise to suitable environments which then survive long enough for life and intelligence to evolve naturally. The mathematics of the Big Bang seem to have achieved a delicate balance since the slightest variation would have led to a dead cosmos either collapsing back in on itself too soon or else exploding outwards with such vigour as to prevent the formation of galaxies. Or if for example the force of gravity were only marginally different there would be no sun-like stars anywhere. Stars of the solar type are very long-lived, burning steadily for many billions of years. They provide the stable conditions which are needed on orbiting planets for the successful evolution of life and mind.

These numerical coincidences may be seen as evidences of design and purpose, though opinions differ as to their significance. A weak form of the anthropic principle, for instance, can dispense with God altogether. It argues that there are an infinite number of other universes in which all possible mathematical relationships are explored. (Rather like the proverbial monkeys on typewriters who given an infinity of time will tap out all the works of Shakespeare.) According to this line of reasoning there is no special value or significance in humanity. The reason we are here and can speculate about the origins and meaning of this universe is that it is the only one, or one of the few, in which the conditions are just right for the emergence of strange creatures who ask questions.

Alternatively this may be the only universe, as supposed by

the strong version of the anthropic principle. Intuitively this feels a simpler and more satisfactory assumption. In this case the many extraordinary coincidences which appear to trim the design of the universe will inevitably become very interesting to us. Although they cannot amount to a proof of intelligent design they do have power to raise the question in an enquiring mind. The significance of coincidence is always open to doubt and we have the added disadvantage that we have nothing with which to compare this universe. We can simply observe that it is remarkably well suited to the task of making people.

The significance of the anthropic principle for the Christian faith is that it highlights the essential interlocking nature of all things. There is great beauty in the unfolding of everything from the first microseconds of the Big Bang right through to the day when the DNA molecule twisted dust into consciousness. All the multitudinous recurring phenomena of creation, daisies, people, stars and galaxies are the result of a basic simplicity, harmony and symmetry of the laws of nature. It is the special insight of physics into reality which reveals this beauty which mathematics can then state in the form of equations. The physicist Richard Feynman has said of this total view of the universe that 'we have to look at the whole structural interconnection of the thing'.[5] It is not enough to reduce the complexity of the universe to its fundamental simplicity. The simplicity and the complexity must be taken together for both tell us some truth about reality. The importance of the anthropic principle for the Christian faith is that it suggests that the universe can be understood in terms of humans or, to make the principle wider, in terms of conscious beings. Human beings may be only one variety of conscious creatures in a cosmos teeming with intelligent life. At this very early stage of our exploration of space the consciousness that emerges from the complexity of the human brain is the highest known to us. Yet it may have more to tell us about material reality than anything else. Everything points to humans, as far as we are concerned, as the key to the laws of nature for there matter realises its true potential. Matter gives birth to the human spirit because it is one of the attributes of matter to do so.

There has always been a suspicion that the universe is merely a brute fact with no inner meaning. The Greek philosopher Heraclitus described it as a flux always changing and likened it to 'a rubbish heap scattered at random'.[6] Today many people avoid thinking too deeply about it fearing that it may be 'ultimately meaningless'. The anthropic principle, rather like the traditional proofs for the existence of God, raises once more the possibility that these suspicions may be wrong. Our sense of the unique value of humanity and of purpose may not be an illusion. Humankind is the measure of the universe and the intuition of the Genesis myth-makers receives the support of science. What science highlights revelation completes.

Whatever tentative conclusions we now draw about the significance and value of the human spirit will depend in some measure upon what we can say with confidence about God. It has been convenient and intuitively satisfying for faith to conceive of God in super-anthropomorphic terms. Michelangelo's famous painting on the ceiling of the Sistine chapel portraying God reaching out to a man is the archetype of this imagery. Many analogies and metaphors in scripture contribute to the same familiar concept. The God of Genesis walked in Eden in the cool of the evening, seeking the ashamed Adam; the creator in Isaiah's monotheistic vision measured the waters of the sea in the hollow of his hand, stretched out the heavens like a curtain and looked down on the inhabitants of the earth as grasshoppers (Is 40:22). This anthropomorphic conception of God is strengthened even more by a superficial and physical interpretation of the belief that humans are made in the image of God. The inevitable consequence is that God has come to be envisaged in a human image.

The new cosmology with its immense time scale requires a new language for theology. Traditional imagery, despite its great poetic power, will need to melt somewhat and be supplemented by other more appropriate terminology and symbolism. A less well known painting by Michelangelo in the Sistine chapel is remarkably prophetic in this respect. In his painting of God separating the darkness from the light, the figure of God is indistinct and wrapped symbolically in a cloud-like swirl of power. Kenneth Clark has compared it appropriately to a twisting nebula where new stars are being born.[7] The creating Spirit is better portrayed by a mysterious 'cloud of unknowing' than by a superman figure.

The modern science writer Paul Davies, author of *God and the New Physics,* has gone so far as to argue that 'science offers a surer path to God than religion'. The simplicity, symmetry and harmony of the basic mathematics which describe the universe is so beautiful that one could well feel that it all points to God. The remarkable way all things seem tailor-made for human creation strengthens the intuition. Not all see it this way however. In a recent book *Creation,* as much a work of literature as it is of science and reminiscent of the prose poem of Lucretius, Peter Atkins an Oxford chemist has argued persuasively otherwise. He dismisses God with an impressive sleight of hand. Most people live their lives shrouded in illusion and baffled by complexity. All religion, he believes, is illusion, it makes some people better people but it is also responsible for a great deal of nastiness. Science meanwhile is revealing the basic simplicity of the universe. Everything — stars, planets, elephants and people — is the consequence of a few basic laws. Eventually, we must suppose, one mathematical equation will be found to account for all that has happened since the Big Bang. Such a universe follows its own simple rules, however marvellously complex and wonderful it may appear on the surface of things, and needs no divine hand to guide it. God becomes an unnecessary hypothesis because there is nothing for God to do. God can then be cut out of the story with 'Occam's razor' that medieval principle which advises that one should 'not multiply entities', or roughly speaking, not accept a complex explanation if a simpler one is at hand. But such neat philosophical surgery is not always necessary or even advisable.

An atheist believes that God is no longer needed as an

explanation. Pre-scientific faith, it is true, depended a great deal upon human ignorance and drew strength from what it did not know. Acts of God abounded; floods, earthquakes, refreshing seasonal rain, even the syphilitic marks on the face of an adulterer, all came directly (it was believed) from the divine hand. As science progressed, more and more of nature was explained without any reference to the supernatural. Some gaps remained in the body of scientific knowledge and theologians often clung to them tenaciously: how did primates acquire intelligence and become people or how did the first living organisms come alive in a dead chemical universe? The questions were frequently loaded in such a way as to presuppose the conclusion that only the activity of God could account for what had happened. This 'God of the gaps', feeding on the hiatuses in our power to explain things was in for a rough time it turned out. Science had no need of God and theology had no respect. Whenever God's existence seemed assured, an alternative, more natural and more scientifically satisfying explanation was uncovered. Even the arch mysteries of how life emerged from molecules or mind from mammals are now seen to be explicable, at least in principle, in terms of the laws of nature. Science dispels mystery whenever it focuses with hard thought and rigorous discipline. The God who sat above the clouds cryptically interfering whenever necessary has been in steady retreat for the last four centuries down a road leading to redundancy and unemployment.

Peter Atkin's conclusion is that such a non-interfering God could be infinitely lazy. The universe can be coherently explained in terms of its fundamental particles, quarks and leptons and the three basic forces which may turn out one day to be aspects of one force. The satisfying simplicity of a grand unified theory (known by its intuitively pleasing acronym as GUTs) rings a death knell for such a divine creator as clearly as Nietzsche's madman who ran with a lamp into the marketplace shouting 'God is dead'. An infinitely lazy creator, argues Atkins, who does not have to interfere from outside at any point need not exist at all. So God disappears in a puff of smoke like a genie back into the void from which God had been conjured by our feverish 'answer-seeking' but 'delusion-creating' mind.

Atkins summarises his own belief in the last paragraphs of a recent book on thermodynamics *The Second Law*:

> We are the children of chaos, and the deep structure of change is decay. At root, there is only corruption, and the unstemmable tide of chaos. Gone is purpose; all that is left is direction. This is the bleakness we have to accept as we peer deeply and dispassionately into the heart of the Universe. Yet, we look around and see beauty, when we look within and experience consciousness, and when we participate in the delights of life, we know in our hearts that the heart of the universe is richer by far. But that is sentiment, and is not what we should know in our minds. Science and the steam engine have a greater nobility. Together they reveal the awesome grandeur of the simplicity of complexity.[8]

How did such an extraordinary state of affairs come about with sentimental creatures delighting in life at the heart of a meaningless cosmos? There is a mystery here of a different order that science will never dispel and before which it can only remain silent. The mystery of existence itself is one aspect. Science can peel back the layers of the onion of matter to reveal simplicity at its heart. Quarks (a deliberately frivolous word gleaned from James Joyce by the mathematician Murray Gell-Mann) and leptons (Greek for 'small coin' – the small change of the universe) are all that the universe is made of. Everything from people to red supergiant stars is made of the same stuff. How did such an elegant universe come to be? Science can only describe the way it *is* whether eternal or finite, and then its task is done. The day will come sooner or later when science *will* be able to take a rest, all explaining done. But questions of a different order will still remain. Suppose the universe is eternal, the Big Bang followed after a multibillion year interval by the Big Crunch only to be repeated again and again, the systole and diastole of a cosmic heart? Why should it be like this when there might not have been a universe at all? Or suppose the more generally accepted assumption is correct that the universe began with a 'seed' and emerged from it for the first time in the Big Bang. The question 'why?' is still relevant. Some people rule such a question as out of order, metaphysical and beyond our answering and therefore nonsensical. Yet we still feel constrained to ask it. If, as faith would have it, the universe is a product of mind, and an intelligent being created it, then it is not unreasonable to suppose that human minds evolved by the process of creation may find an answer. The answer, in fact, is more likely to lie in our questioning than in the chemistry that science describes. (Though our appreciation of the fundamental simplicity and beauty of that chemistry should not be left out of the answer.) The wakeful experience of the questioning human spirit must tell us something about the cosmic process of which it is the most remarkable product. The tentative nature of the language in which our answers may be couched should not be allowed to undermine the very proper status of the question.

Traditional Christian theology claims that God caused the universe to be out of nothing, *ex nihilo*. Though even Thomas Aquinas acknowledged that his cosmological argument for the existence of God as First Cause would still stand in the case of an eternal cosmos as proposed by the philosopher Aristotle. The divine creator is then conceived as the underlying cause of the eternal series of temporal causes. However, several attempts have been made this century to demonstrate that the universe could have come out of nothing of its own accord.

Mythologies in the past developed their own marvellous poetry of nothingness. Teutonic mythology begins with this image of the void from which the All-Father made something emerge:

In the Ages, when naught else was, there yawned in space a vast and empty gulf called Ginnunga-gap. Length it had, and breadth immeasurable, and there was depth beyond comprehension. No shore was there, nor cooling wave; for there was yet no sea, and the earth was not made nor the heavens above. There in the gulf was the beginning of things. There time first dawned . . .[9]

The Shastika Indians of California speculated that Old Mole (id-i-dok), a huge animal, heaved the world into existence like a molehill out of nothing. The Pitt River Indians attributed it all to Coyote the wild dog who scratched it up out of the void.[10] In Genesis there is a hint of the same sort of language, a poetry of primeval emptiness, when it says that 'the earth was a formless void . . . and God's spirit hovered over the water' (Gen 1:2, JB). An even older allusion to nothingness survives in the English word 'abyss'. The root of the word is related to the Sumerian creator God Absu. The mystery of 'something' where there might be 'nothing' has entertained human imagination since before writing was invented.

Modern cosmologists face exactly the same problem, it seems to me, as their spiritual forebears. We do not have the language to speak about the relationship between existence and non-existence. Atkins explores quantum mechanics for an answer. In an energy field, matter and antimatter can spontaneously appear as it were out of nothing (but only 'as it were' because what in fact happens is that energy is transformed into particles of matter obeying Einstein's law that $E = mc^2$). A proton and an antiproton may materialise for a moment and then within the briefest fraction of a microsecond cancel one another out again. By analogy the universe itself could have emerged from nothing. Rather as one can reverse the equation which states that 'plus one' and 'minus one' equal zero and say that nothing equals 'plus one' and 'minus one', so one can argue that the total energy of the universe is ultimately zero. If a bank manager were to allow me to open two bank accounts one containing a hundred pounds and the other containing an overdraft of a hundred pounds then for a time, out of nothing I have something.

Deft as these analogies may appear they both founder on the same problem as that faced by the analogy of matter and antimatter spontaneously materialising in 'empty' space. They all presuppose the existence of *something* whether numbers or bank managers or simply energy fields. They still tell us nothing about why such an odd state of affairs should be the case at all. How is it that there is a universe based upon such elegant quantum possibilities? Atkins' own account of the bizarre fact of existence is as poetic as that of any of the prescientific myth-makers of the past. In *The Creation* he writes about this three-dimensional, time-bound universe emerging from 'a dust of unstructured points'. Even more enigmatically he suggests that 'somehow matter must have been created out of something resembling nothing'. This cosmos is just one of the possibilities latent in nothing, he argues. It was an opportunity for 'something' with a good survival value which is the reason why it has been around long enough to throw up people who can observe and question it. But considered in its totality this enduring evolving universe which generates elephants and people ultimately cancels out to nothing with the finality of a neatly balanced equation.

In the dark night of nothingness a brilliant flower bloomed and fell: a cascade of beauty falling from nowhere to nowhere. This is the strangely evocative picture that Atkins paints for us, a surrealist vision of the cosmos. But it is an affirmation of faith depending every bit as much on the sentiment of the atheist as belief in God does on the sentiment of the Christian. The traditional proofs for the existence of God have been seen, for a long time, to be on shaky ground; but then, initially, their purpose was not so much to prove that God exists but to establish God's relationship with creation; they were all formulated in an age which did not really doubt God's existence. In St. Anselm's words they were 'faith seeking understanding'. Today their only value may be that they 'bring God back into the discussion', the judgement of Hans Küng, Roman Catholic Professor of Dogmatic and Ecumenical Studies of the University of Tübingen.[11] An atheist today is on exactly the same unsure ground if he insists that the universe came out of nothing, meaninglessly and of its own accord. He can no more, on the grounds of logic, establish his No-God in that gap than a theologian can smuggle her God in as creator.

Theology has already accepted the fact that only inadequate pictures of God depend upon what we do not know. Religion is not frantically seeking the dark corners of science to prop up its faith. Today our knowledge of God will be mediated much more by the insights of poets, saints and physicists into the way reality is perceived rather than depending upon our ignorance. Behind the first microseconds of the Big Bang lies an impenetrable veil. It is a boundary of space and time, an edge, an ultimate horizon. Beyond that we face what Stephen Hawking, Cambridge mathematician and astrophysicist, has called a 'principle of ignorance'.[12] It is as meaningless to discuss either mathematically or physically what lies before the Big Bang as it is to ask what one would find a mile north of the north pole. Any speculation about where the singularity came from, the seed of the Big Bang, is essentially metaphysical. When Christian theologians adopted the Big Bang theory and rejected the steady state model of the universe (discarded since by the astrophysical community) they were reacting more romantically than logically to the new discoveries of science. The picture suited their way of thinking. For a moment it seemed as though science had uncovered the secret of the universe and disclosed the work of God. Scientific explanations came to an end, providing a heaven-sent opportunity to ascribe the universe to a transcendent cause – God. But Kant two hundred years previously had already shown that we cannot rationally speak about transcendent causes. By definition what is transcendent lies beyond the realm of scientific theory. Nothing is proved.

Much depends on what sort of emphasis we put upon our view of things. Atkins, stating the atheist position, judges statements of faith to be products of sentiment. They are not what we should know with our minds which, with cool rationality, will see that they are illusions conjured up in a purposeless universe. But here we must be careful not to be led astray by the undertones of words. Atkins is using the word 'sentiment' in a very eighteenth-century way; 'a thought coloured or proceeding from emotion', 'as opposed to reason' (OED). It implies a shallowness as in 'sentimental', whereby one is apt to be swayed by sentiment or addicted to indulgence in superficial emotion. In earlier medieval usage the word had none of these trivial, shallow connotations. Sentiment was 'personal experience', 'one's own feeling or mental attitude'. In other words, the way I see things, my own *vision*.

We are, for the most part, badly tuned instruments, and the varying ways we see our lives and our environment are a mixture of true and false slants. This does not mean that every vision of the world, every product of sentiment, is wrong or that we will never attain a true outlook onto reality. The conviction of faith is that God has caused us to live in an autonomous universe. The great beauty of science's description of the world is that everything can be accounted for by the working of natural law. The process of creation and evolution is beautiful and complete in itself. But why this delicious unwinding of reality from the Big Bang to the awakening consciousness of the human spirit? If persons are the highest product of the process then it is reasonable to give a high place to personal experience. The sentiment based upon generations of insight, inspiration and revelation which views the cosmos as purposeful and people as receptacles for grace is of no less value than the sentiment of an atheist who views the universe as a surrealistic bloom in the void. Each, Christian and atheist, has a way of looking at the strange facts of our existence.

Deep down the argument between the atheist and the Christian is about two things; about what philosophers have called 'picture preference'[13] and about how one evaluates the sentiments of the human spirit. The militant rationalist who practises extreme reductionism will see the cosmos in wonderfully simple terms. The mathematical simplicity, symmetry and harmony will be sufficient to satisfy the spirit since it can account for the whole rich jungle of reality which has emerged through time. The Christian on the other hand, putting more emphasis upon the complexity and intricate interlocking order of nature (though without denying its beautiful, fundamental simplicity), will see it in holistic terms reckoning that the reductionist approach is not enough. Truth is revealed as much through complexity as it is through simplicity, otherwise you might as well reduce architecture to bricks, symphonies to sounds, and poetry to the alphabet. The spiritual insights of psalmists and poets (themselves amongst the finest products of chemistry) are potentially just as true as those of the mathematician or the physicist. Zen Buddhism bears ample witness to the fact that enlightenment may be attained at many levels and in many ways that are not just mental in a narrow sense. As in archery or flute playing, the whole person – head, heart and body – is involved. The scientist does herself a disservice if she thinks of herself only as a scientist or if she rejects the possibility that truth may be perceived through the symbols of faith.

It is really only one image of God that contemporary cosmology has knocked off its throne. Michelangelo's God sitting above the world and leaning over now and then to poke a finger into creation is a cartoon that has had its day. The concept had already come in for a lot of criticism after the eighteenth-century deists turned God into a super cosmic artisan who constructed a perfect universe like one of those polished brass mechanical orrerys now only found in museums. The descriptive power of science had so delighted theologians with the intricate details of the workings of the heavens above the nature on earth that in their own way they honoured God by imagining God to be a perfect artisan, the precise clockmaker of William Paley. It was poets of the nineteenth century, Wordsworth, Tennyson and others, who appreciated the danger of this picture. A flawless universe only required God to wind it up in the beginning and then God could step back and let it run on its own. The infinitely lazy God of Peter Atkins, the God who need not exist at all is the logical descendant of the God of the deists.

The divine creator conceived by contemporary faith has to be intimately involved with this creation and not a distant supernatural being. There can be no resting yet for God on a seventh day unless we conceive of God as working and resting at the same time. The world is not finished and the world is full of pain. It is far from being a perfect place, except insofar as it is the ideal environment for independent souls to grow in freedom. If we believe in the creative activity of God then we have to interpret the first statement in Genesis in the present tense. *Now* is the beginning and God *is* creating the heavens and the earth. Contemporary theology looks for new language and tends to speak of the creator as the 'divine ground' of the universe. The phrase is not new. Paul Tillich borrowed it from Hinduism and it can also be found in the meditations of the medieval saint Julian of Norwich.[14] This creating spirit, the eternal presence, is the field in which the reactions of fundamental particles and forces have their being and weave their work. Creating and sustaining from moment to moment, the Spirit makes possible the emergence from matter of the spiritual dimension. Everything, every photon and every quark, every sparrow on the housetop and every hair of the head is absolutely dependent upon God's sustaining loving power.

The image of God to emerge in us was already potentially present in the first microseconds of the Big Bang. Through the aeons the creating Spirit has acted like a careful caring midwife preparing for and attending the birth of a new creature able to reflect and comprehend eternity. The human spirit rooted in the temporal qualifies for the eternal. Raised from dust it shares the life of its maker.

Cosmological theory is incomplete if it does not contain an account of the life that blossoms naturally in the cosmos and has human consciousness and creativity as its finest bloom. Science clarifies the picture of the expanding universe and reveals something of its vastness, its antiquity and its awe-inspiring beauty. Faith may then contemplate all this as the incredibly beautiful work of God who determined the fundamental laws, the nature of the basic particles and forces, knowing that given time and free play they would produce life and finally people. It is a universe made to make the human person.

4. EVOLUTION

In 1861 Charles Darwin wrote a book on orchids, which caused considerably less stir than the volume *On the Origin of Species* published just two years previously. It bore the rather ungainly title *On the various contrivances by which British and foreign orchids are fertilised by insects and on the good effects of intercrossing*. The *Literary Churchman* of the day welcomed it as a work which gave praise to God acknowledging that in an indirect way it was saying 'O Lord, how manifold are thy works'. Asa Gray, Darwin's contemporary, wrote that if the work on orchids had appeared before the *Origin* the author 'would have been canonised rather than anathematised by the theologians'.[1]

So why is it that Darwin's theory of evolution by natural selection has been such a stumbling block to so many Christians? Churchmen of the nineteenth century, such as Bishop Wilberforce, are not the only ones who have been bitterly averse to the idea. Today, more than a hundred years later, there are large pressure groups particularly in the USA who want to restore the idea that God created in six days as a piece of genuine science. 'Creationists', as they have come to be called, attempted to put 'creation science' onto the syllabus of schools in the state of Arkansas as recently as 1983.[2] Although defeated in law the attempt may well be repeated in other states. Sadly the debate has polarised into a false dichotomy. An unrealistic choice is offered, one which prejudices the issues involved. Either God created in six days, it is argued, or else evolution happened without God's help. But even before Darwin's day many Christians had accepted as reasonable the notion that God might create by means of evolution. As early as the fourth century the great theologian St. Augustine had written 'in the beginning were created only germs or causes of the forms of life which were afterwards to be developed in gradual course'.[3] The dispute still rages however. For the lay person and the non biologist the whole thing can be very confusing.

The theory of evolution transformed the Christian world view, probably more than any other scientific concept. Gone is the idea of an original perfection, an ideal world spoilt by a man's sin in the garden of Eden. Process and change replaced the static picture. The kindergarten image of God creating in one go all creatures great and small as though for a nursery Noah's ark is discarded. A rolling picture takes its place, of experimental life forms, trial and error, gradual modification or sudden mutations and developments, of random extinctions, failures, oddities and triumphs. The complex web of life that wraps itself around Planet Earth has evolved as an interrelating and interdependent system, transforming itself and its own environment over immense periods of time. The place of people in this new vision of an evolving ecosystem is less obviously prestigious. The dividing line between people and brute beasts is blurred. The neat mythical pattern of Genesis which rounds everything off in six days is replaced by a three billion year process without any end in view. The picture of God, who is responsible for this process, has had to change too. God has to be involved not just 'in the beginning', but now, intimately creating moment by moment, using the laws of physics, chemistry and biology and the opportunities offered by chance. Design is not imposed upon creation but emerges with the process, and God is Living Spirit not a retired Architect.

The theory of evolution has a great contribution to make to the Christian faith, but it still has to overcome in some quarters the mistaken assumption that it is contrary to Christian teaching. Why do some people find it so unsettling and such a threat to the Christian gospel? It is important that we understand some of the reasons for opposition to the theory before investigating its positive value. Both the Christian who is a scientist and also the theologian must address themselves to these fears.

A changing world is certainly an unsettling world but the most disturbing aspect of the theory of natural selection is that it seems at first sight to dispense with God. Life creates itself. Creatures evolved to fill the gaps and niches in the environment because they were there to be filled. The logical working of the laws of nature was enough to determine their various characteristics and abilities, their colour, height, hairiness, and so forth. There is no need for science to invoke a creator to account for all the remarkable adaptations that make the network of nature function so well. The long bill of the hummingbird so suitable for probing the trumpets of tropical blooms for nectar was selected by circumstances. (The flowers in this example are beneficiaries of the arrangement too. They developed their shapes by coevolution with the hummingbirds which pollinate them.) The camouflage of the tiger, the thick coat of the polar bear, the streamlined shape of the dolphin, are all the result of the witless pressures of the environment and the needs of the creature. Even beauty in nature, the song of the woodlark, the tail of the peacock, are the result of evolution following its own laws. God, it seems, is out of a job. A divine Designer is an unnecessary hypothesis.

Possibly the most anxiety-provoking element in the theory is that evolution by natural selection has no goal and no aim. It has no direction. It is intrinsically and essentially purposeless, as haphazard as the falling of dice. There was no intention in nature, for example, no prior plan, to adapt the dolphin or the whale to a watery environment. While their mammalian cousins survived on land circumstances were such that the ancestors of today's whales and dolphins found it easier to get on in the sea. Bred by circumstance, they evolved and took up the opportunity of an environment rich in food and relatively free of predators. Similar pressures were a major factor in giving shape to all other organisms. A significant feature of this aimless process is that it depends not on forward planning but on the hazards of chance and accident.

Changes in the climate, the drifting of continents and separation of land masses, or the variations in the food supply are just some of the chance circumstances which help to determine the changing shape or function of any individual plant or creature. It is the haphazardness and unpredictability of evolution humankind is just one of the accidental products of the whole aimless process. Not only is God unnecessary but then life would be different, and according to the theory of evolution mankind is just one of the accidental products of the whole aimless process. Not only is God unnecessary but the theory of evolution suggests that people are simply by-products of an unfolding purposeless accident. The thought can be deeply disturbing.

The mythical story of creation in Genesis sets out quite deliberately to generate a different impression. God is responsible for every detail. People are at the centre of the stage. The meaning of creation is seen in terms of man and woman made in the image of God. The sun and the moon are lights to illuminate their days and nights. Plants are there for them to eat and animals to be named and controlled. God made the world and gave it to these Lords of the Earth to dominate and rule. There are no doubts here about the value and significance of humans or of their moral responsibilities and it is the moral clarity of this picture which is so appealing. A deep-seated fear today is that life is ultimately absurd, an accidental bloom in a cold and alien universe. This world just happened to be, and life developed on it in a random way. An implication of living in a godless world is that the moral sense may be undermined. This ought not to be the case, perhaps, but it is. Anything goes, nothing in the end really matters. Morality becomes convention or private self-protection, but conventions can change and successful selfishness might be more realistic than loving your neighbour. Evolution is a struggle and the fittest survive. The same egocentric law becomes the basis of morality for individuals and nations.

None of these fears is justified, I am convinced. It is true that the world faces in the immediate future immense moral dilemmas. It is also true that many people – practical atheists

– live lives, some happily, some disastrously, without reference to God. It is undoubtedly the case too that the sense of meaninglessness, that life is mad and nothing seems to really matter, is very widespread. But clinging to the Genesis myth as though it were a scientific statement will not in the end resolve any of these problems. Indeed it will hinder, I believe, any attempt to confront them realistically. Genesis is not an alternative to evolution, but understood as myth it can make a unique contribution to the world view provided by science. It can illuminate and bring deeper meaning to the scientific insight that human consciousness is an emergent quality of matter disclosed by evolution. The universe, through the very nature of its fundamental laws, was destined to produce observers given time. The very existence of men and women may be the key to understanding the universe at all levels. We should not discard the Genesis story but see what light its inner convictions bring to the world we inhabit.

An account of the universe will be incomplete in a major respect if it does not include a description of its most extraordinary product: life. The evolution of life on earth is intrinsically connected to the nature of the cosmos itself. Butterflies and the Big Bang are both aspects of one unfolding order; supernovas and human nativities have deep links. By what genius did the atoms which fell to earth from the sun and the stars weave themselves into such intricate patterns; a filigree of order and design? The glorious gold of buttercups and the mind of Beethoven are each of them the products of necessity, time and chance, woven from atoms which in themselves are purposeless and unconscious. The mystery of evolution is revealed in an even more splendid epiphany when its mechanisms are described at the scientific level. Darwinism, Mendelian genetics, microbiology and all that is currently being discovered about DNA contribute to a vision of the most wonderful phenomenon in the universe, the weaving of a web of life.

Darwin's great contribution to the theory of evolution was the law of natural selection. There is a simple and delightful logic about it. It was this theory which he revealed in 1859 and which was discovered simultaneously by A. R. Wallace. Evolution as such was not a new concept. Decades before the publication of *On the Origin of Species by Natural Selection* to give it its full title, the general idea of evolution was assumed by many biologists, Erasmus Darwin, Charles' grandfather, being one of them. The problem in Darwin's day was to uncover the mechanism of the process. How did evolution happen? Lamark had already suggested one possible solution, since discredited, based upon acquired characteristics. (Any feature a parent improves is passed on to its children. But no amount of stretching by the giraffe will produce long-necked offspring.) Darwin's and Wallace's great and original discovery was to demonstrate that evolution by natural selection was a logical consequence of three generalisations, themselves dependent on careful research and observation. Accept the generalisations and the conclusion unarguably follows.

Firstly there is variety within any generation of a species; some individuals are taller, hairier, faster, longer beaked, thicker skinned or more alert than their fellows. They are not all exactly alike, whether fish, plants, mice or men. Secondly, parents tend to pass on their own special characteristics (not those acquired during their lifetime) to their children. The selective breeding of race horses or show dogs depends on this principle. Thirdly, when there is a food shortage, a change in climate, a new predator or circumstances are adverse for some other reason then only some of that species will survive. Some will be able to cope with the new conditions others will not, depending on how well their individual characteristics adapt them to the change. This third generalisation became clear to Darwin as a result of his reading the works of his contemporary Malthus who wrote about the pressure on human populations. Natural selection is an unavoidable deductive consequence of these three premises and we should then expect to see evolution happen whenever there is life in the sea, the sky, on land or indeed anywhere in the universe. The law applies equally well at the level of the tiny amoeba as it does for elephants.

Some of Darwin's followers applied the theory of natural selection vigorously to the whole of evolution and allowed no other mechanism for change. Darwin himself did not espouse such a hyperselectionist point of view. According to this extreme interpretation of the principle, every item of nature, every organ, every wing, every feather, every behavioural trait had been selected for its useful purpose. Nothing was useless and if some feature seemed superficially to serve no useful purpose then it was our ignorance not the theory that was to blame. Such a strict and exclusive application of the principle is of course impervious to disproof. A theory that rejects all exceptions does not make much progress. Darwin himself was never as extreme. He speculated for instance about the role sexual attraction played in selecting colourful features in mates and he acknowledged the possibility that sometimes changes could happen which were simply useless. He protested that he was often misrepresented on this matter. In the introduction to later editions of *The Origin of Species* he wrote 'I am convinced that natural selection has been the main but not the exclusive means of modification. This has been of no avail. Great is the power of steady misrepresentation.' Recent work by Neo-Darwinists, who have integrated Mendelian genetics and discoveries about DNA, the genetic molecule, into evolutionary biology (the Modern Synthesis) suggests other mechanisms besides natural selection. Subtle genetic variations may take place within the whole gene pool of a species causing change without any external circumstantial pressure. This is known as 'genetic drift'. Another similar mechanism has been called 'molecular drive'.

There have been many exciting developments in evolutionary biology since Darwin's day. Ideas are subject to change every bit as much as living organisms. Darwin's original theory itself has evolved, been corrected and added to year by year. One of the great scientific revolutions of the century was the elucidation of the double helical structure of DNA in 1953 by Nobel prize winners James Watson and Francis Crick.

Little was known in the nineteenth century about the

mechanism by which inherited characteristics were transmitted from parents to their children. Darwin had proposed his own highly speculative theory which he called pangenesis. It involves tiny messenger particles, gemmules, which travelled from every part of the parent's body to the sperm and the egg. They were supposed to pass on detailed information about shape, size, colour, etc., which would define the characteristics of the subsequent child. The theory was completely misconceived. The transmission of inherited characteristics remained an unresolved problem for Darwin. Nor was anything known in his day about the source of random variations within a generation upon which the principle of natural selection worked. The discovery of DNA and the elucidation of its structure was able to answer both these problems.

The remarkable DNA molecule is basically a double string of smaller molecules. The arrangement of these smaller units, nucleotides, constitutes the code which records all the characteristics of the parent organism. The two threads of molecules twist together in a long spring-like helix. They are closely linked because each nucleotide connects with an opposite number in the other string. Thus they form pairs. One of the marvellous things about this chemical code is that it is an alphabet containing only four letters. All the threads of twisting DNA in every living organism on earth are composed of only four types of nucleotide, which are themselves built from common atoms such as hydrogen, nitrogen and carbon. That simple alphabet can define and then produce the material to make a mouse or a man, an orchid or an oak tree. It also defines behavioural characteristics, the mouse-catching instinct of the cat, the suicidal behaviour of the lemming, the migratory instinct of the swallow. It accounts for self-centred behaviour in humans and may even be responsible for acts of altruism (bombing your enemy *and* loving your neighbour are very likely to be codified in the DNA of *Homo sapiens*).

One of the most extraordinary discoveries of molecular biology is that every cell in your body contains all the information needed (encyclopaedic in quantity) to make an exact replica of you. It is all there stored in the threads of DNA.

This bizarre fact has become the theme of many science fiction stories. One of the intriguing problems yet to be adequately unravelled by science is how each cell of the body knows what sort of cell it has to be. How does it know which 'page' to turn to in the encyclopaedia of information stored in the DNA so as to become a brain cell, a piece of muscle, a finger tip or an eyelid? A complete description of *this* process will be invaluable for understanding cancer and why maverick cells behave the way they do.

Mutations, changes in the genetic code, due to incorrect copying (the DNA equivalent of Freudian slips and typists' errors) or the effect of radiation or some mechanism of random shuffling, account for the variations amongst individuals which makes natural selection an effective principle. Without these mutations there would probably be no change, and thus no evolution.

A lot has been added to Darwin's work, but there is nothing in the discoveries of molecular biologists which undermines his basic theory. There have been no modifications or additions to evolutionary biology which we can confidently say would not have delighted Darwin had he lived long enough to see them. This needs to be emphasised because the popular press and television often give the impression when reporting disputes in this field of research, that Darwin has been disproved or that evolution itself is being seriously questioned. Nothing of the kind is happening.

Darwin used Herbert Spencer's phrase 'the survival of the fittest' to underscore what happens when natural selection 'steers' evolution — if one can speak about the 'steering' of a process which in its details is essentially aimless. Two examples will illustrate this point. The first is of a small but measurable change in finches of the Galapagos Islands. (Darwin had visited these Pacific islands himself in the *Beagle* and observed the finches in some detail.) In the five years between 1975 and 1980 a drought killed off much of a variety of grass there. Finches that fed on its small seeds had to turn to another food source. The only plants available had larger harder seeds and only those birds with the toughest bills could cope with them. Many finches died and only those with larger beaks survived to breed. In the five year period the average bill size of this particular sort of bird increased by ten percent.

The second example, a famous one, shows more dramatic change but over a longer period of time.[4] The peppered moth, *Biston betularia*, once had light patterns on its wings. This camouflage made the moth almost undetectable to predatory birds if it was resting on a tree trunk. One in a thousand of these moths was a dark coloured variety. This genetic variation was there in the gene pool but only surfaced occasionally. With the industrial revolution tree trunks in the English Midlands became blackened with soot from factory chimneys. The light coloured moths now stood out clearly to be picked off by birds, while those with black wings merged into the background. Within a century the dark forms had replaced the lighter variety, being best fit to survive in the changed environment. These are only small examples but evolutionary theory claims that much greater changes in shape and colour are the result of the same process. In a period of sixty million years all the mammals, elephants and antelopes, whales and weasels, mice and men have evolved from a small shrew-like creature with grasping hands that lived in trees.[5] Go back 300 million years and reptiles share the same ancestor with mammals. Two and a half billion years ago a primitive organism was the common ancestor of both woman and wheat. All this as a consequence of the purposeless laws of evolution. Change just happens. The fittest survive and breed. Evolution produced monkeys and humans without intending to.

The traditional Christian concept of God as designer clearly has to be rethought. None of the species began as they are now. Indeed ninety-nine percent of all the species that have inhabited planet earth are now extinct. God's involvement in the process of creation is considerably more subtle than theology had imagined in the past. A subtlety that has led some to the conclusion that God has faded from the scene altogether like the smile of the Cheshire cat. S. J. Gould in his collection of essays *The Panda's Thumb*,[6] makes an entertaining jibe at the traditional concept of God as designer. Nature contains so many odd modifications and awkward solutions that it looks as though it is full of afterthoughts and temporary accommodations and is generally something of a Heath Robinson job. There is no evidence, he argues, for a perfect designer. An intelligent God would never have designed the panda, for example, the way it is.

The panda, beloved of children and the media lives (when it is not in zoos) in remote parts of China where its staple diet is bamboo. It has evolved an extra thumb which aids its bamboo stripping needs. A bone in the wrist called the radial sesamoid has become enlarged and as a consequence developed its own complex set of muscles and nerves. Other related, non-bamboo eating bears do not have this odd feature and from the point of view of evolutionary theory it is a neat demonstration that change happens. Adaptation results from the pressure of needs and circumstances. Gould sees the panda's thumb as proof that there cannot be a God who designed everything and fitted each creature to its ecological niche. It all smacks too much of the second thoughts of an amateur. Like the tiny atrophied legs of the boa constrictor visible in the skeleton but normally hidden beneath the snake's skin, or the useless appendix in the human being, the panda's thumb is one more oddity, this time useful, which suggests a less than perfect plan in the mind of the creator. 'Odd arrangements and funny solutions are the proof of evolution – paths that a sensible God would never tread but that a natural process constrained by history follows perforce.'

The sixth digit on the pad of a sedentary Chinese animal is not the only example Gould uses to discredit the idea of an ideal omnipotent creator. Orchids which fascinated Charles Darwin and took up much of his time, have evolved a whole variety of breathtakingly beautiful methods for trapping flies so that they can cross-fertilise. They have formed an alliance with insects by manufacturing from the common components of ordinary flowers intricate devices to attract them. A conventional petal is modified so that it becomes an inviting trap, providing a landing pad and a spring mechanism for holding the insect long enough to dust it with pollen. Darwin's book on the subject, mentioned earlier, is a compendium of these subtle devices. They are living evidences that orchids must have evolved from ordinary less exotic flowers and Gould sees them as further proof that there is no ideal creator: 'Orchids were not made by an ideal engineer. They are jury rigged from a limited set of available components.'

How does God relate to evolution then? Gould's judgement that orchids and pandas prove that there was no designer rests upon a common preconception of the divine, one which many theologians discarded long ago. The seventeenth and eighteenth centuries saw the highpoint of a piece of philosophy known as the teleological argument. (From the Greek telos for 'end' implying 'purpose'). It argued that every biological species was specifically designed to serve its own needs and all the fine adaptations of nature proved the existence of an intelligent designer. One of the most famous treaties on the subject, *Natural Theology* by William Paley used the celebrated analogy of 'clockwork'.[7] The intricate interlocking mechanism of a clock exists because of the clockmaker. Similarly the regular mechanistic movements of the heavenly bodies and the remarkable adaptations of nature all argue for there being a divine 'clockmaker'. This well-worked argument collapsed like a house of cards when the wind of change blew through biology.

One of the first things a Christian may have to do if evolutionary biology is to be allowed to make a contribution to faith is to let go some of the traditional ideas about God. Thank God for the panda's thumb, faith may say, for with a flick of the wrist it releases us from an outmoded static image of God. Theology itself has evolved and the idea of a divine clockmaker setting an unchanging mechanistic universe in motion then letting it run on its own, is now part of history. The real God got up from the throne in the sky where the deists put God in the eighteenth century and walked off beyond the horizon long ago.[8] The task of faith is to pursue the living God who calls us to follow.

Attempts have been made to update Paley's clockwork analogy by restating it in terms of twentieth century engineering. This might solve the problem of how God relates to evolution. Many computerised factories today use computers and robots. These machines themselves make the machines such as cars which are the product of that particular factory. If twentieth century people can make machine making machines then by analogy one could argue that nature is a machine-making machine.[9] This allows Darwin to explain by natural selection how the various adaptations of organisms came about without reference to God, and leaves God responsible for the laws that run the whole show. The solution has a weakness however. It is in danger of pushing God one stage further away from the living creatures that inhabit this world. And that is not consonant with other insights that surface regularly in the biblical tradition that God is intimately involved with every detail of creation. God knows every sparrow on the roof top and has numbered every hair on your head. Theology which incorporates evolution must be true to this belief, and then it may find in the end that the creator is much closer and more available to us than we ever dreamed possible.

A major challenge to faith will be to incorporate 'chance' and 'accident' into a theology of creation. The words carry heavy emotional overtones of arbitrariness, suggesting that the various products of life on earth are ultimately the result of haphazard forces. If evolution proceeds by the unguided accidents of circumstance then people might never have emerged on earth. It is an angst-provoking thought. Our planet could have been covered with luxuriant jungles of foliage infested with insects but with no one there ever to see it. Go further back down the evolutionary tree and the phenomenon of life itself might not have emerged from the molecules of the young earth. This universe could have remained a lonely realm of dead chemistry.

Did God leave it all to chance then? Is God a gambler and did God just *hope* the world might evolve creatures to be divine offspring? The possibility is disturbing. Intuitively it feels unsatisfactory. Professor H. N. V. Temperley, mathematician and author of *A scientist who believes in God*[10] reveals some of this unease. A particular issue about which he is unhappy is the mechanism by which the first living cells appeared on earth. Were they the result of a natural process or did God have to breath life into the first living creatures? 'My conclusions, which, for religious reasons, I do not particularly like, is that what must surely have been the first step in the evolution of the primitive soup *could* have occurred *by accident* (my italics, AF) and that it is therefore unnecessary to invoke either divine creation or the appearance of protolife from interstellar space to explain the initial step.'[11] This conclusion with its theological unease about a process of creation which involves chance appeared in a *New Scientist* article, 'Could life have happened by accident?'

It is our fear of purposelessness that makes us so uneasy about giving chance and accident a role in creation. Without purpose our lives become meaningless and the spectre of Heraclitus' universe, 'a heap of rubbish scattered at random', begins to haunt us. This fear and the threat of extinction by thermonuclear war are perhaps the two most unsettling elements in the psyche of people today. Both fears are the result of scientific research and both bring us to the edge of nothingness. We might not be here if evolution had followed another route, and we may cease to be here quite soon if humankind is bent on self-destruction.

Accidents must be seen to have a positive role in our theology of creation by evolution. One fruitful line of thought uses the analogy of a game. A game has rules and constraints but it also contains an element of chance. If it were only rules it would be no game, but chance makes it come alive. God's act of creation is a continuous thing like the playing of a game; all the moves cannot be calculated and predetermined from the beginning. The game has to be played to find what opportunities are offered by chance. There was not a major event 'in the beginning' when God created everything; as creating Spirit, God is the ground and cause of the universe making things 'be' day by day, instant by instant. God is generating an open-ended universe governed by law and chance. It is not a closed mechanical system running in predestined grooves. The creating Spirit knows that given time, the laws of physics, chemistry and biology (which at root are all the same thing), and the opportunities offered by chance, evolution is bound to follow a route leading towards molecular complexity, life and consciousness. The day to day working of the laws has no aim yet the general outcome is intended. The same process can be described in these two ways without contradiction: at one level evolution is aimless, while at another level it has purpose.

This paradox of purpose and purposelessness can be illustrated by a simple example. If I pour, aimlessly and haphazardly, handfuls of marbles into a large mixing bowl, they quite naturally without any intervention take up a honeycomb like pattern of hexagonal groups. The marbles fall by chance in any number of ways but they are then constrained and ordered by the laws of physical geometry. The marbles have no aim or purpose and do not know that they will make a pattern, but I do. Similarly the purpose of the creating Spirit is to produce people. The Spirit is the field in which Darwinian evolution happens. It generates and continues to create the scene, the fundamental laws of nature, the material of the universe and the opportunities offered by chance which together constitute the patient process of evolution. A pattern has been threading its way unconsciously through the chemistry of the earth for billions of years and through the universe for four times as long. It did not know what it was doing until it wove humankind into the fabric.

The concept of two levels of truth makes it possible for the biologist to speak about evolution as aimless while the theologian can say quite justifiably that it has purpose. It is not necessary to reject the ideas of goals or intentions altogether, as the philosopher Jacques Monod (1910–76) has done, arguing uncompromisingly that evolution is solely the product of blind chance. Chance can have its own fruitful part to play in an overall plan. Divine accidents are not meaningless.

'God does not play dice', said Albert Einstein, founder

figure of the new physics. He was contemplating the enigma of quantum indeterminacy and rejecting the idea that God's creation might contain an element of chance. It made the relationship between God and people too arbitrary. With game theory we find that there is another alternative. 'Certainly God *does* play dice' argues Manfred Eigen, German theologian biophysicist and Nobel prize winner, 'but he also follows the rules of the game.'[12] It is in the working of these two aspects of evolution that the creating Spirit, Lord of Law and Lord of Chance, provides both meaning and freedom in our world. Once we grasp the nature of the process then we need no longer fear that chance implies ultimate purposelessness. God's intentions are not compromised by the aimlessness of evolution. The truly remarkable thing is that the laws are so simple. As we saw when considering cosmology in chapter three all the forces of nature may one day be shown to be variations of just one superforce united in a grand Unified Theory. Energy transformed into matter manifests a similar simplicity. Out of this basic symmetry and simplicity emerges everything both in and under the sun, from protons to people. The human spirit is created by this deep play of God quite naturally and logically, law and chance are all that are needed for the evolution of creatures made in the divine image.

Man and woman are made in the image of God, but God is creator and so by implication we are invited too to become co-creators. We can shape the world around us and direct our own futures. Life on earth has always been changing or evolving. The only new element, though it is a major revolution, is that humanity can now consciously direct and steer that change. We have always had the opportunity to shape our world to some extent by the various moral and non-moral decisions that we make. This generation is seeing new and undreamt of possibilities opening up before it. Conscious goals are now part of the selective process which leads to change. What began as a relatively unimportant activity, the modification of our domestic pets by cross-breeding has developed into a world-saving industry. Selective breeding which produced fantailed pigeons and bulldogs now develops short stemmed, heavy eared grains that crop three times a year and so keep pace with the demands of the world's fast expanding population. New strains of heavy cropping rice; disease resistant potatoes; animals with a high percentage of lean meat that fatten quickly; protein from fast-growing fungus: the list is almost endless. The very survival of the human race may depend upon our ability to harness evolution in our own interests.

An uneasy transition is happening as we become masters not victims of circumstance and of the subsequent process of evolution. It is an inevitable destiny and part of the divine plan that we discover the image of God within us and take up some of the responsibility for shaping the world. It is also fraught with risks and dangers and the moral dilemmas that already face us are immense. We are by no means in control of, or cognisant of all the repercussions of our new-found powers. Genetic engineering and the newly acquired techniques for recombinant DNA experiments open up a road leading to paradise or hell depending on your hopes or fears.

The Christian cannot avoid facing these issues. Although there are no simple answers nevertheless with hard thought and an honest appraisal of the insights of both science and of Christianity a framework of faith may be forged which will offer good guidance. Firstly God is not absent from the process. It may be a useful metaphor to describe God as a designer and allowing that the laws of nature are responsible for all life on earth, to argue that God designed a 'machine-making machine'. The danger of such an image is that it tends to leave God detached, alienated from the process. God did not set in motion a self-creating world and then leave it to its own devices. God is the eternal presence, the creating Spirit causing everything to be, moment by moment. God is the Field in which the game is played, intimately close to every detail of the process. This insight into God's involvement with creation is already there in the Old Testament. The psalmist expanded on the theme

O Lord, thou hast searched me, and known me. Thou knowest my downsitting and mine uprising, thou understandest my thought afar off . . . Whither shall I go from thy spirit? or whither shall I flee from thy presence? If I ascend up into heaven, thou art there: If I make my bed in hell, behold, thou art there. If I take the wings of the morning, and dwell in the uttermost parts of the sea; even there shall thy hand lead me, and thy right hand shall hold me. (Psalm 139: 1–2, 7–10, AV)

Lord of scientific law and Lord of chance, God is not detached from the process of evolution. But neither is humanity. The second conviction that the Christian will hold to is that our life (the human phenomenon) is an extension of the life of the world. In pauline phraseology the whole of creation has groaned to see the emergence of offspring (Rom. 8:22). The divine purpose is that people should be products of a material process. We are now in the remarkable position, to an increasing extent, of taking control of the world. That does not imply however that we can now deny our roots and selfishly exploit the rest of nature. We should never allow ourselves to become alienated from the process from which we have emerged. The Christian framework of faith will have a high regard for the material world and for other emerging creatures. As we began to distinguish ourselves and our destiny from the rest of creation it was natural to think in terms of dominating our environment. As our spirituality becomes more sophisticated it is now appropriate that we think rather of coexisting with nature and not just in terms of conquering it. We can only take up the responsibilities of co-creating with God if we truly love the world.

Nothing has been said yet about fossils, and this may seem an odd omission. Many people are under a mistaken impression that the theory of evolution depends on the fossil record, that it stands or falls on the accuracy or interpretation of that evidence. It does not. There are many arguments which work together to support the claim that species evolve. There is the intrinsic logic of the situation already discussed, that variation, heredity and population pressure will inevitably lead to change. Then there is experimental evidence provided by artificial selection; dogs, cattle, race-horses, rice, chrysanthemums all bear witness to the success of breeders and cultivators. By creating our own goals, we deliberately do what nature had done arbitrarily. Our choices are the selective pressures which steer change in whatever species we want to improve for our own ends. Supporting all this evidence, nature has produced its own examples of evolution in the hundred year period that separates us from the publication of *The Origin of Species by Natural Selection* – the peppered moth, the Galapagos finches and many others. Evidence of evolution is with us too in other ways. There are 'living fossils', vestigial features found in animals that are 'withered' as it were through lack of use; the legs of the boa constrictor, the wings of the kiwi. Wherever the biologist looks, there are indisputable signs of change. Even if all creatures crumbled immediately at death leaving not a single fossil in the world, the biologist would still have to conclude that evolution is a fact of nature.

'Creation science', the theory which says that the Genesis myth should be interpreted as literal history denies evolution and relies heavily on questioning the evidence from fossils. One of the leading protagonists of this view D. T. Gish entitled a book *Evolution: the fossils say no!*[13] Christians who find the theory of evolution not only acceptable but even offering positive insights that support their faith, must handle this issue with great care. It arouses strong feelings. Disagreements should be faced with compassion and understanding not arrogant rejection. Some commentators write off the creation science position as intellectually bankrupt, a bigoted stance by blinkered reactionaries. They see it as a fraud and the attempt to have it taught in American schools as an unforgivable attempt by a religious pressure group to indoctrinate children. (The issue did become so inflated that in the 1980 presidential election candidates had to make their position on the matter clear. President Reagan declared himself doubtful about the evidence for evolution.) The important issue, often forgotten, is why the creationist interpretation of Genesis has such appeal.

The literal interpretation of scripture, as has been discussed earlier, can be traced back to the sixteenth century, and could be argued to be a heresy which resulted from the Reformation. It is an irony that the Genesis myth got turned into science as science itself began to explain how the world works. The continuing appeal of the fundamentalist position is that it contains clear straightforward claims. God made the world and God made people. The purpose and meaning of life is to give glory to God and to fulfil the moral purposes God has for us. Let go the story as a piece of literal history and, it is feared, the claims about God and people will evaporate, leaving us without moral guidance, lost and meaningless, our privileged position in nature undermined. The spectre of a godless uncaring world haunts the fundamentalist and until that is exorcised he or she will cling quite understandably, if misguidedly, to a literal interpretation of Genesis.

The creationist position was made clear in the 1983 Arkansas trial when the attempt was made to put the six day creation theory into the syllabus of schools' science teaching.[14] The First Amendment of the American constitution protects schools from any threat of religious coercion so that religious instruction is technically banned. The creation science theory claims that the universe and life was all created suddenly out of nothing, relatively recently. Not all fundamentalists insist on the accuracy of the precise calculation by archbishop James Ussher that the world began at 9 a.m. on 23 October in the year 4004 BC. For most creationists 'recently' means within the last ten thousand years. They deny that mutation and natural selection are sufficient to account for the development of all living creatures from a single organism and claim that changes only happen within narrowly fixed limits. God created the original species of animals and plants and the ancestry of humans and apes is separate. Any evidence to the contrary, such as the fossil record and the apparent age of the various layers of the earth's rocks is explained by the occurrence of a devastating worldwide flood; the story of Noah. This last element of the theory has been called 'Catastrophism'.

The majority of biblical scholars today propose that the Noah story is an ancient folk tale incorporated into Genesis to make a theological point. There is archaeological evidence of large-scale flooding in Mesopotamia the undoubted source of the story and there is a close parallel in the Epic of Gilgamesh in which Utnaphishtim is the hero who escapes the deluge.[15] The question being asked and answered in the Bible is 'Why if God saw that the world he had made was good did he then attempt to destroy it?' It is the question 'why?' which is always asked when faith is confronted by evil. Fundamentalists interpret the story literally as a worldwide flood and consequently believe that archaeology should find evidence of Noah's ark on the top of the 16,946-foot peak of Mount Ararat in eastern Turkey. No less a 'modern' person than James Irwin, an astronaut who walked on the moon, has led a team to make such a search. This seems to forget that if all the water locked up in the ice caps were melted and if all the water in the air were to fall to the ground as rain the sea level of the world would not rise more than a few hundred metres at the very most.

Catastrophism suggests that during the flood the earth's surface was buckled and twisted providing those misleading clues which led the geologist Lyell, Darwin's contemporary, to conclude mistakenly, they believe, that the various rock strata were the evidence of the ancient earth. Geologists have been misinterpreting the signs ever since, according to creationists. Thus catastrophism explains how sea shells can be found in the rocks at the tops of mountains. Fossilisation happened almost instantaneously in the disrupted post-diluvian landscape. Layers of coal were created in a few moments under extreme pressure and not in the millions of years most of us have been led to believe. Even the progressive nature of the fossil record can be explained by this scenario. The lowest rock layers contain fossils of the simpler creatures such as trilobites, while middle layers contain dinosaur bones. Fossils of mammals and remains of early humans appear near the surface. This apparent progression leads atheist thinkers to conclude that evolutionary development has taken place, so the fundamentalists argue. Catastrophist theory offers another explanation. When the flood water rose in the days of Noah all the living creatures about to be drowned were sorted out in a rapid forty day struggle. Primitive shelled creatures immediately sank and stayed at the bottom. Large animals like dinosaurs and reptiles lumbered their way up the sides of mountains only to be overtaken by the rising tide a few days later. The human species scrambled to high places and were overtaken by the flood last of all. Only Noah and his loaded ark escaped completely to float away privileged and elect into a new world.

'Creation science' is a misnomer since there is nothing that science can investigate which will either support or refute the claim. It is a fixed idea imposed on the world, and the evidence will always be made to fit. Even leading creationists acknowledge that it cannot really be presented as an alternative scientific theory. 'We do not know how God created, what processes were used, for God used processes that are not now operating anywhere in the natural universe. This is why we refer to divine creation as special creation. We cannot discover by scientific investigation anything about the creative processes used by God' wrote Gish in *Evolution: the fossils say no!* John Whitcomb Jr and Henry Morris write in the same forcefully anti-scientific vein in their book *The Genesis Flood*. 'It is thus quite plain that the processes used by God in creation were utterly different from the processes which now operate in the universe! The creation was a unique period entirely incommensurate with the present world.'[16] The call for establishing the Genesis myth as scientific theory to be taught in the schools of Arkansas, failed by their own admission.

It is not really the facts which are in dispute between creationists and evolutionary biologists. The debate is frustrating to both sides because it is to do with a way of looking, rather like some of those visual illusions produced by the artist M. C. Escher. The patterns on the page look like a flight of malevolent bats until suddenly you look again. The same page and the same patterns now appear quite different. This time what you see is a flock of white doves flying in the opposite direction. Sometimes it is difficult with such a picture to switch from one way of viewing it to the other. The evolutionary biologist and the fundamentalist are stuck with their respective interpretive pictures of reality. How do we know which is right, and which makes best sense of God and the world? There are, I believe, strong theological reasons for preferring the evolutionary picture.

God could, of course, do anything. That presumably is the implication of the traditional teaching that he is omniscient and omnipotent. If it had suited him he could have made the world in the year 4004 BC, complete, finished and out of nothing. He could even, if he had so wished, made the world just one hour ago. Consider this possibility for a moment. At his command adults, adolescents and infants appeared on the scene already dressed or lying in bed, well fed or hungry. Roads, railways, and aeroplanes; houses, cinemas and factories, mansions and slums, supermarkets, orchards and vegetable patches were all willed into existence, into solid working reality. He made libraries too, full of books about histories he had conjured up to give a background to this present moment. Old manuscripts, documents and ancient looking earth works completed the illusion of a past. God gave memories to the people he had made just one hour ago, and he linked these people with loves and remembered quarrels. He set these people off in various directions some to walk home and unwittingly meet their families for the first time, others to continue reading books they had never in fact read, others to leave prisons for a free world they had never known, remembering crimes they had never committed. He put photographs on mantlepieces and in wallets and left old newspapers fresh from his mind strewn across chairs and tables and wrapping fish in the freezer. He gave people common recollections of great wars, Independence Day celebrations and world cup matches. All these things he put onto a planet which he had produced out of the void like a conjurer palming a ping pong ball out of the air, and he set this planet adrift in immense space to orbit a gigantic sun. Distant galaxies and exploding stars he made to fill the skies so that people on earth could go on looking forever and never see the edge of everything; needing no edge to the universe, he refrained from creating one. All this happened sixty minutes ago. Sixty-one minutes ago there was nothing.

The point about this bizarre speculation is that it is quite clearly metaphysical. Metaphysics is beyond scientific investigation. No experiment, research or observation by the most meticulous or imaginative scientist could dig up a fact to disprove this metaphysical theory. The combined evidence of every historian, biologist or astronomer and the witness of every private memory would be of no avail in the attempt to disprove such a fanciful claim.

The creationist position is similar, I believe. God could have created the world in six days, less than ten thousand years ago. God could have given Adam and Eve navels (a matter of heated dispute and speculation in the nineteenth century), the severed links with wombs they had never known; put trees in the garden of Eden with tree rings registering years of growth that had never been; manufactured fossils in the earth and crumpled the landscape with a great catastrophe. No amount of reference to the real world will disprove such an *idée fixe* if its protagonist has other overwhelming reasons for clinging to it.

Creationists counter this accusation by pointing out that exactly the same can be said of evolutionary biology. What discoveries, they ask, would count as evidence to falsify and to disprove the theory of evolution? Too many biologists, it could be argued, assume that evolution is true without questioning it. Their method is always to explain facts, even the most awkward ones by recourse to the theory rather than have the facts undermine it and subject the theory itself to the test. Karl Popper, the philosopher of science, unintentionally added fuel to this accusation by debating the philosophical status of the theory of evolution. A theory is only truly scientific if it can be tested, that is to say subjected to an experiment which could prove it wrong. Like everyone else scientists can be emotionally biased and crave for their own theories to be right. Awkward exceptions are ignored or 'adjusted'. But this is the wrong view of science, Popper points out. Scientists should always be exposing their theories to rigorous tests, trying to question them, looking for the exception that will *disprove* the rule or at least modify it. He called this the falsification principle. On these philosophical grounds evolution cannot technically be called a scientific theory. It is more a way of looking at the world, a conceptual tool for interpreting biology.

In consequence Popper suggested that the theory of evolution should be called a 'metaphysical research programme'.[17] He did not deny the value of this way of seeing the world or even its truth but was simply pointing out, as a philosopher, the peculiar status of the theory. Unfortunately many of the general public and creationists in particular thought that Darwinism was being questioned at a fundamental level. It was as 'a metaphysical research programme' that evolution came to be described in the Natural History Museum in South Kensington. The media then took up the creationist case quite uncritically and gave support to the idea that Darwinism was unscientific and therefore by implication untrue.

Perhaps a better way of describing evolution is that used by Wilma George in her biography of Darwin. It is a 'working hypothesis', and as a working hypothesis has an even more secure status in our description of physical reality than does a scientific theory. It is so coherently reasonable on so many grounds that to continue to question it as one might a scientific theory would be to waste time and energy. The concept of evolutionary change does have a firm hold on the minds of the majority of biologists, not because a religious or private interest is at stake, but because it makes such good sense. It is not a perverse concept based only on the wish to disprove the existence of God supported by the dubious evidence of a few fossils. The theory is the result of an accumulation of insights from a variety of fields, biogeography, genetics, artificial selection and palaeontology all supported by the conclusions of deductive logic. The fossil record plays a small contributory role, though with improved techniques even this is gaining significance. It would be as foolish to treat the theory of evolution as a temporary and falsifiable explanation of nature as it would to keep questioning whether the laws of physics are the same on the moon or in New York as they are in London.

The only real arguments among biologists are about *how* evolution happens. The basic theory makes such good sense that it no longer needs to be questioned. Regrettably, however, the various arguments within the field of biology are interpreted by some creationists to mean that scientists are questioning the theory itself. Often television and the newspapers take up and then further misrepresent the issues involved. We have already noted the stir caused by the philosophical debate about whether to call the theory of evolution a 'scientific theory' or a 'metaphysical research programme'. Other issues too are prone to casual misunderstanding and misrepresentation. When discussing the role played by the sun and the stars in the evolution of life (chapter two) we looked at the very real difficulties science has in explaining how life appeared as an emergent quality from megamolecules. These difficulties are not proof that the scientific approach is wrong or that the general theory of natural selection is inapplicable at that juncture of creation. A very similar and intriguing problem will be discussed in the next chapter when we will consider the status of mind as an emergent quality of complex brains. Disagreements amongst anthropologists and zoologists about the mechanism of evolution at this point are disagreements about *how*, not *whether*, an evolutionary process works.

Perhaps the most misunderstood discussion in evolutionary biology is the issue of speciation. The creationist claim is that God created all the species of plants, insects, birds and animals at one time. They allow that small changes may have taken place within species but deny that evolution could cause completely new species to emerge. The issue of speciation is very alive today and there are plenty of opportunities for Neo-Darwinists to disagree with Darwin. But, again, because he was wrong about details does not mean that he was wrong in general. So for instance Darwin believed that evolution was a slow and gradual process, rather as he imagined the growth of mountains to be. His colleague T. H. Huxley warned him that in this instance he was loading himself with an unnecessary difficulty. Darwin's faith that 'Natura non facit saltum' (Nature does not make leaps) is today doubted by many Neo-Darwinists who hold that long periods of changeless equilibrium are punctuated by bursts of rapid evolutionary development.

Creationists always argued that the fossil record supported their case that evolution cannot produce new species because it is very difficult to find evidence of any missing link creatures, half way houses between one species and another. If evolution has been a gradual process of adaptation and change with all the varied life forms fanning out from a few or even one basic tree of life, then the fossil record should show these changes like the still frames of a film. In fact the fossils reveal nothing so straightforward as that. There are gaps in the record and there are long periods covering millions of years when creatures seem to have changed very little, if at all, and there are very few examples of fossils that trace for us the evolutionary links between species. In other words it tells us very little about the way a species branches away from another. There are no almost-horses and no almost-cats. The 'six day creationist' who would like to retain the nursery image that God made cats and horses in the beginning just as they are today, will find support for his picture from this negative evidence.

S. J. Gould and N. Eldredge have proposed a solution to this enigma which is entirely consistent with Darwinian theory and accounts for the observed facts. They call it 'punctuated equilibrium'.[18] This theory accounts for the development of a new species, which they call 'speciation', by proposing that it happens rapidly in a small isolated group. A small interbreeding group separated from the rest of its species by a mountain range, river or sea may then 'go its own way' in new surroundings and, without the control and check of the gene pool of the original species, develop into something quite new. Such a small group facing totally new conditions could undergo a significant change in form very quickly, that is to say within a hundred thousand years. The newly-evolved type, if it is well adapted to the environment, will then become stable in that new form and breed unchanged for many millions of years. Compared to this later steady existence in equilibrium when there is little change in the structure of the creature, the process of speciation is rapid and may account for only one percent of the newly-evolved creature's history. Furthermore, it will have happened within a small interbreeding group whose numbers would be insignificant compared to the great population of their descendants once the new form has become stabilised. Consequently the chances of finding fossil remains of the early developing form of the species are remote. They will not be distributed widely but will appear in some isolated pocket and will be comparatively few in number. By analogy, if it were not for a few keen collectors and an occasional museum, a visitor from space might conclude that the motor car had appeared on earth in its efficient mass produced form. The early trial models are now few and far between and have only survived as collectors' items. Only a very few of the earliest examples of horseless carriages show their obvious links with that older form of transport.

Punctuated equilibrium is not an argument against Darwinism but a modification of it. Gould and Eldredge have suggested that evolution did not happen as a gradual ascent, but in fits and starts, explosions of new forms being followed by long periods of equilibrium. It is the latter which predominate in the fossil record, as one would expect, and so we are presented with a somewhat unbalanced picture. Natural selection is still the principal mechanism of evolution.

In fact the fossil record is somewhat better today than many people, particularly those opposed to the theory of evolution, suppose. There are now quite a number of examples of creatures in fossil form which are the missing links we have been looking for between species. There is a great difference between mammals and reptiles visible to even the most casual observer for instance; yet we now have the collected fossils of a whole spectrum of creatures who became extinct some 230 million years ago which provide a link in the fossil evidence between reptiles and mammals by being their common ancestors. Another example known to Darwin is the fossil of archaeopteryx. The skeleton of this winged creature reveals it to have been half bird, half dinosaur, and it is a clear example of the transformation that can happen through time of one type of creature into another. The terrifying pterodactyl, so popular in children's books, is related to the garden robin.

There are also living examples of speciation happening now. When a species is examined in detail over a wide geographical area it is found to vary slightly from place to place. Sometimes a species has spread right round the world with interesting results. The British herring gull is connected with a number of slightly varying interbreeding populations, distributed in these latitudes right around the North Pole.[19] The gull's appearance varies slightly as you move west from Britain across the Atlantic and North America. By the time you reach Siberia the bird begins to look more like a black backed gull. When you reach Europe and Britain the bird is indeed now a true black backed gull which does not interbreed with the herring gull. It is a small detail but clear evidence that speciation can occur through the ordinary mechanism of evolution.

Why should God so confuse the descendants of humankind with so many false clues? If the creationists are right then God seems to have led a great many honest God-fearing thinkers quite deliberately up the wrong garden path.

The image of God evoked by the creationist interpretation of the world and of scripture is one of the weakest elements in their thesis. The highest attribute God gave us humans is reason. It seems odd that God should then put us into a world in which through the rigorous application of that reason we are led irrevocably to draw the wrong conclusions about its nature. What possible purpose could God have in leading people astray with so many evidences of this being an evolving world if the contrary is true?

The fundamentalism of the creationist is based, I believe, on a sad misunderstanding of the way in which scripture is inspired. God did not dictate a piece of scientific history to the authors of Genesis. God inspired them to see certain truths about God, us and the world. The Roman Catholic church's current view of scripture is that it is not the words which are directly inspired but the authors who wrote them. The Holy Spirit inspires rather than dictates. A decree of the 1965 Second Vatican Council contains these words, '. . . the interpreter of sacred scripture, in order to see clearly what God wanted to communicate to us should carefully investigate what meaning the sacred writers really intended and what God wanted to manifest by means of their words'. The Holy Spirit inspires rather than dictates, and in the words of the Protestant theologian Leonard Hodgson we should always be asking 'What must the truth have been, and be, if people who thought and wrote as they did, put it like that?'[20] Our reading of scripture should be just as inspired and guided as was its writing, and we need to develop in our day a confident doctrine of the inspiration of the reader.

The creationist makes the mistake of equating value judgements with scientific facts, of confusing different levels of truth and misinterpreting the nature of myth. He or she is misled by a modern literalistic interpretation of scripture. The intention of the creationist to restore value to humans and honour to God in contemporary society is wholly good, but the method, I believe, is profoundly misguided, and in the long run generates far more problems than it solves.

The theory of evolution gives glory to God and contains within itself something like prophecy. It has great potential for contributing to a theological view of the universe and the discovery of meaning through a deeper understanding of the spiritual dimension of matter. Its advocates are not atheistic apostles of a godless materialism. Far from dispensing with religious faith it can provide a new way of looking at the mystery of creation. The universe is unfolding some of its secrets and revealing its meaning to those with eyes to see.

One of the first theologians to have an insight into the spiritual implications of evolution was the Jesuit Teilhard de Chardin.[21] Many Christian thinkers had already realised the beauty of the process and acknowledged God as Lord of evolution, but they had not taken the further step and combined the new scientific world view with traditional theology. Teilhard accomplished this with a coherent vision of creation. It was his genius to look at both evolution and Christianity with new eyes and to see the world as a God-filled and God-directed process. Evolution, a stubborn scientific theory which would not go away and which until then had been married awkwardly to scripture became the source of new opportunities for faith.

The Roman Catholic theologian Hans Küng has paid tribute to the part Teilhard de Chardin played in creating a new harmony between science and theology. In his book *Does God Exist?* he wrote:

> It was only in the course of time and in fact only from Teilhard de Chardin onward that theologians increasingly began to observe what *new opportunities for the old faith* were offered in particular by the evolutive understanding of the world: opportunities for
> a deeper understanding of *God* – not above or outside but in the midst of the world and its evolution;
> a deeper understanding of *creation* – not as contrary to evolution but as making evolution possible;
> a deeper understanding of the special position of *man* – not as independent of the animal, of his history, of his behavior, but as a being of body and mind in his unique relationship with God.[22]

Teilhard de Chardin died in 1955. He had spent much of his life travelling in Asia doing palaeontological research. Throughout his life he was forbidden to publish any of his speculative writings because of their possible heretical implications. After his death his work exploded dramatically onto the theological scene and made a lasting impression on both Catholic and Protestant theology. He had sown the powerfully fertile seeds of a new concept. What is remarkable is that he generated his vision in such isolation. Although he was allowed by his religious order to travel the world as a palaeontologist he had no opportunity to see how his ideas would be received by the public, or to test those ideas against a critical audience, answering criticism and query.

Obedient to the disciplines of the two currents in his life, science and theology, Teilhard felt an urge to develop a vision which would unite them. The evolutionary development of life from molecules through simple cells to complex brains, from dead matter to conscious spirit has followed an aimless path. Or so it seems at the level of scientific description. Yet with hindsight, looking back down the tree of evolution from its branches to its roots, Teilhard could see a pattern and a purpose emerging. He put great emphasis upon the importance of 'seeing' the significance of the world process. Just as we may see the point of one of Jesus' parables or be blind to it, so the concept of evolution in all its breathtaking detail may speak to us of a deeper meaning. Or as the events of Israelite history, the exodus or the exile could be interpreted by contemporary prophets as the activity of God so those with scientific training may read the signs of the processes of nature. In each case the facts of the matter are the same, whether the details of a parable or the events of history. It is the way of *seeing* them that differs. Teilhard contemplated the progress of evolution and saw in it a pattern, an arrow pointing upwards from the past through the present and into the future. This arrow follows the evolutionary path of ever-increasing complexity and convergence and traces an ascension. With vision we may see this arrow pointing towards God, 'the Omega point' in Teilhard's phraseology, the End towards which evolution slowly makes its way. Just as surely as texts in the Old Testament were read by early Christians as pointing to the coming of the Messiah, so the scientific account of evolution may be seen to contain its own message.

The most significant evolutionary development on earth since the growth of that layer of life encircling the globe known as the biosphere (analogous to the atmosphere or the lithosphere) is the emergence of thought. We are all born into a vast system of communications and ideas and we breathe our cultures like the air. Teilhard called this network of thought the noösphere, (from the Greek for 'thought' – 'nous'). The inherited thinking of society shapes us as people and without it we would be gibbering monosyllabic savages. Sometimes Teilhard wrote of the mantle of thought which enfolds the world as though it has a life of its own quite apart from the individuals who do the thinking. It is difficult to know whether he was writing mystically or metaphorically when he developed this idea. Whichever is the case, the significance for him of the noösphere is that its appearance on earth is a major stage in evolution's journey towards God.

Christianity was then identified by Teilhard as the emergence of a new mode of consciousness within the noösphere, the birth of a new order. This is of immense importance to our understanding of the role Christianity is destined to play in human society. If his vision of the natural world is true then Christianity is not merely a spiritual epiphenomenon of society. It reveals the way to a new stage in the growth of humanity. Pluralist societies have tended to reduce religions to private opinions, comforting but optional. But the clear message of Christian revelation united to evolutionary theory is that life on earth faces a new frontier. Perhaps that accounts for the stark alternative offered in the pages of the New Testament; either we find eternal life or we perish.

At the heart of Teilhard's vision is the figure of Jesus Christ. In Christ all that can be said about God comes together in one focus. The Omega point, the goal towards which evolution drifts and in which it will have its fulfilment has already been united to the physical process of evolution in a creative act of extraordinary mystical unity. Made flesh it dwelt among us. This new level of evolution involves an even more radical leap forward than the one when people emerged from apes or when life emerged from molecules. Jesus Christ is a descendant of the great evolutionary family tree which has been unfolding around the planet for three billion years. He is a product of nature. He is also from God. In this mysterious union of the human with the divine, of the world described by science with the kingdom of God, lies the hope of humankind.

The theory of evolution invites religious people to embrace the material world with some enthusiasm. Spirituality is not a means for escaping the pains and tribulations of the physical but a way of illuminating them. God evolves us from the chemistry of the earth, and we are made of dust. Christianity is not the only religious tradition to begin to see the significance of this point. While Teilhard de Chardin was travelling the world as a palaeontologist and developing his vision of evolution as a great sea drift making its way towards God, a similar process was unfolding in the mind of an Indian philosopher, Aurobindo Ghose.[23] I was introduced to Aurobindo's work in 1968 when I visited Dr. Radhakrishnan, statesman and philosopher, at his home in Madras. I was particularly interested to know what impact science had had upon Hindu religious thinking. Dr. Radhakrishnan quoted the Taittiriya Upanishad written in the sixth century BC which describes the evolution of spirit from matter.[24] It ends with a song and a cheer that we, who are spiritual beings, are made of physical food. 'You see,' observed Radhakrishnan, offering me peanuts and pure lemon juice from a silver tray, 'the idea that spirit is dependent on matter and emerges from it through the intermediary stages of life and consciousness has been in Hindu thought from the beginning, the theory of evolution is no surprise to us.'

Aurobindo contemplated Darwin's theory of natural selection in conjuction with the Upanishads. Although he was less well informed scientifically than Teilhard, there are passages in his mammoth work *The Life Divine* which are remarkably similar to the Jesuit's. Both philosophers find their faith being kindled by a pattern and direction which they see in the process of evolution. For both philosophers the emergence of spirit contains a revelation of meaning. But whereas Teilhard imagined the further course of evolution to run through Christ and to follow an axis that lies through Rome and the sacramental system of Catholic Christianity, Aurobindo saw it emerging in communities of 'Knowing people' whom he called gnostics. His own Ashram at Pondicherry with its emphasis upon shared work study and meditation is one such community. Hindu philosophy is able to add one refinement (or complication) to the concept of the world as a process which is not available to the Christian: reincarnation. It was anathematised as a heresy by Christianity in the fourth century. According to this doctrine every individual makes the whole evolutionary journey through thousands of lives from the lowest living creature up to a human. All souls travel a great migration from the dark shadows of ignorant matter to God the ultimate goal of their pilgrimage. For Aurobindo God is Brahman; Teilhard, as a Christian, symbolised it as the Omega point, the focus towards which everything in the universe is constrained to converge.

Science without faith offers us a bizarre picture of the world. By an ultimately absurd accident, without meaning or purpose, a universe appears from nowhere, and evolves. Given time, the atoms of the earth link themselves together in such a way as to produce consciousness. Like dust devils, vortices of activity whipped into temporary spinning existence by the desert wind, human beings emerge from the earth through the driving pressures of evolution. People briefly walk the planet, examine the dust from which they emerged and gaze at the sky. They collapse back into the dust as meaninglessly as they emerged from it. Each of us can only look around for a moment before we die.

The prophetic insight of Teilhard was to see more in evolution than simply the functioning of the laws of physics, chemistry and biology. The apparently aimless course of evolution begins to glow with significance when viewed with the questions raised by faith. Prophets of the Old Testament could read a pattern and meaning in the disasters and successes of their history; political events spoke to them of God.

Christian faith today may contemplate evolutionary biology in the same way, and find mystery. The description of the world provided by science is not changed by one atom when interpreted by the eyes of faith, it simply takes on a new and fundamental significance. Running through everything there is a tide, which rising up, bears us towards God. In 'The spiritual power of matter' Teilhard had this to say:

> Son of man, bathe yourself in the ocean of matter; plunge into it where it is deepest and most violent; struggle in its currents and drink of its waters. For it cradled you long ago, in your preconscious existence; and it is that ocean that will raise you up to God.[25]

The oddity of human wakefulness in a material universe is not a bizarre, absurd brute fact without meaning or purpose. It is the birth the universe has unwittingly awaited for billions of years. The mystery of mind emerging from matter will be the subject of the next chapter.

5.   THE HUMAN SPIRIT OR 'DOES MIND MATTER?'

The mystery of mind emerging from matter is one of the greatest wonders of the world. But what *is* mind? Is it a mere side effect of chemistry, an electro-chemical 'fizz', as it were, of the brain? Do only human beings have minds? Can minds exist independently of bodies? As I write these sentences, and the reader reads, we each of us know a mind directly – our own. But we can never know anyone else's. Our hearts, lungs, livers and guts all work without consciousness and successfully perform complex chemistry day in and day out. A new dimension, however, is revealed in our brains; self-awareness. It is here that beauty is perceived, that imagination bestows on us a creative response to the future, that values are held and moral judgements made. A number of words and concepts cluster together as we attempt to make sense of the human spirit – mind, soul, self-awareness, consciousness. The value and status we give to the human spirit must be central to any attempt we make to understand the universe. It will radically affect the attitude we adopt to our own lives.

The study of astronomy has inevitably led us to turn our eyes from the stars above down to the earth beneath. When an astronomer investigates the heavens the more remarkable phenomenon sits at the eyepiece end of the telescope. He or she who works in the observatory, questioning, theorising, or interpreting the data from cameras or computers is the truly extraordinary manifestation of matter, far more significant than the spectrum of a receding galaxy or regular beat of a pulsar. Science itself finally directs our attention to the scientist, the individual with a questioning mind. 'What is man, that thou art mindful of him? and the son of man, that thou visitest him?' asked the psalmist (Ps 8:4, AV). The question is with us again.

Minds may not be unique to Planet Earth. There could well be other intelligent creatures on other worlds who look out into the cosmos trying to unravel its mysteries. The possibility is taken seriously in many quarters and a research programme named SETI (Search for Extra-Terrestrial Intelligence) has been set up in America motivated by the expectant hope of locating other technological civilisations in the immediate locality of our galaxy. The search is conducted by radio telescope on the reasonable assumption that any technologically-developed species would use radio waves for communication at some point in their career.[1] A number of rough but quantifiable assumptions are made in estimating the chances of our being able to 'listen in' to the communication network of an alien race. A cautious approach will limit the possibility of intelligent life to planets orbiting sun-like stars. Only a small percentage of stars in our galaxy are of the solar type. A fraction of these, though it may be the majority, will have a family of planets. A small proportion of these planets will offer hospitable environments for the evolution of organisms (one out of nine in the case of our own solar system), of these only a few may have seen the emergence of intelligent life. A final salutary calculation to go into the equation involves estimating, on the basis of no experience, how long an intelligent civilisation may survive without immolating itself once it has discovered nuclear energy. A conservative conclusion based upon very cautious estimates and assumptions reveals nevertheless a remarkable figure. As many as half a million civilised technological communities may reasonably be supposed to inhabit this galaxy alone. A new radio-communicating species could on this basis be appearing in the Milky Way once every decade. As astronomers turn their radio telescopes towards local stars and tune in to wavelengths, the human race has been likened to a family who have just had a telephone installed. We are waiting for it to ring.

There are intriguing implications here for theology. The image of God, already potentially present in matter, has been realised on earth in humans. If life is spread throughout the universe, whether it began in billions of Darwin's 'small warm ponds' or in the reaches of interstellar space on cold dust particles, will it eventually generate creatures in millions of locations who are made in the image of God? The discovery one day of an intelligent alien race somewhere 'out there' (and even the theory which predicts such a discovery), helps to liberate the image of God from its anthropomorphic mould. We are not created in God's image in the sense that God has ten fingers and toes, arms, legs, head, two eyes and a nose. The image is neither male nor female despite all artistic 'evidence' to the contrary. The bearded Ancient of Days is merely a cartoon device, a symbol for minds unable to cope with abstract thought. The true image of God is the spirit of inwardness which emerges from our human chemistry. It is the mind which with the heart can act freely and knowingly with love and imagination. It is a wakefulness which has the opportunity, denied to earlier stages of evolution, to respond creatively and graciously to its environment. A new factor, the liberated mind, begins to share with God the task of creation.

Alien creatures who swim deep in extraterrestrial oceans or who fly high in clouds beneath remote suns may claim, as confidently as Christians do of themselves, that they are made in the image of God. If rationality with a life of spiritual awareness has awakened in them then they will have every right to do so. Intelligent beings, whose chemistry is based on silicon-oxygen chains rather than carbon, crawling across deserts in other worlds may tell themselves stories similar to the creation myths of Genesis. Wherever chemistry has spun organisms who have minds, and who can make value judgements and act with justice, the image of God will begin to be realised. Their inner lives will bring them close to their creator.

Mind was a possibility latent in the Big Bang. Given the simple laws of nature, time and the opportunities offered by chance, consciousness was bound to emerge one day and there may have been innumerable routes to its realisation. The anthropic principle, that hunch discussed earlier, which suggests that the universe was made to make people, should perhaps be renamed the mind or the consciousness principle. Biochemical evolution may have produced questioning creatures in all manner of shapes and sizes throughout the galaxies. Alternatively it might be the case that whenever organisms with conscious minds are evolved their physical form will be like, or very similar to, that of *Homo sapiens*. There may be an inner logic to the physical structure of humanity just as all snow crystals have hexagonal shapes. In the animal kingdom on earth 'convergent evolution' is a frequent phenomenon. Nature keeps throwing up the same shapes in response to the pressures and requirements of a particular niche. Large creatures which swim in the sea, for instance, whether they be mammals (whales), fish (sharks) or saurians (ichthyosaurs) have evolved a common shape to cope with that environment; a streamlined body with stabilising fins and a strong swimming limb at the back. And the eye of the octopus developed along a separate pathway of evolution from the mammalian human eye. The results are remarkably similar as optimum solutions converge on one advantageous pattern.

Convergent evolution may just conceivably have produced minds in humanoid creatures throughout the cosmos. This possiblity, however, should not make us lose sight of the truth that the image of God is not to be understood in terms of physical shape. Spiritual inwardness transcends the material human form. It is the mystery for which science has no language and can only be approached obliquely through myth, metaphor, poetry and symbol.

Suppose, one day soon, the cosmic phone *does* ring and we find, as we suspect, that we are not alone in the universe. Will the Christian church initiate a mission to convert the aliens and will it assume that the birth of God's 'only Son' was unique to earth? First of all the communication problems will be formidable, assuming that we find ways to understand their language. Even if life is spread through the galaxy with the sort of density suggested earlier that still means that we can only expect one other civilisation within a radius of 300 light years of the solar system. Recall that radio waves travel at the speed of light. A single question and answer would take 600 years if it was with an inhabited world 300 light years away. Of course we may be lucky and find that we have closer neighbours. But having made contact we may then find that it is *we* who are in need of conversion. Many extraterrestrial civilisations may be vastly more ancient than ours and spiritually far more advanced. St. Athanasius' tenet that 'God became man in order that man might become God'[2] will need extrapolation. The word may have become flesh aeons before Jesus' ancestors had even woken up to being a chosen people. What other Christmas nativity scenes may there have been already in other epochs beneath other stars? My own guess is that like convergent evolution there will be convergent theologies throughout the galaxy. We can look forward to a fruitful dialogue.

A more immediate problem for theology and biology is how we may speak about the difference between us and animals. Or, since we are animals then what is it that sets us apart from other animals? The traditional answer was simple. People have souls, animals do not. People can sin, repent, be forgiven, discover immortal life, go to heaven and so forth. Animals are amoral with no sense of right or wrong and when they die, they die. No resurrection awaits them after death. In the Christian creed the community of saints was never expected, except by the most romantically minded and those with a naive view of paradise, to include dogs, cats or chickens. The trouble with this traditional view, however, was that it depended on a picture of the world in which the human was a creation apart from the animal kingdom. It proposed a radical discontinuity between us and all other creatures. Our rational minds and creative spirits, which could begin to take responsibility for our own actions, set us apart and above the elephant, the horse and the cow.

Now the boundary between human and animal has become blurred. We are primates with far more in common genetically with our chimpanzee cousins than we have dividing us. (99.5% of the DNA, the molecular code which defines our physical and behavioural characteristics, is exactly the same.[3]) An unbroken continuity of changing forms links us through our primate ancestors back to a tree-shrew-like creature of the Cretaceous era seventy million years ago. The oldest evidence we have so far of our hominid ancestors is the footprints of a male and female who walked in the ash of a volcano after rain, three and a half million years ago. Did souls enter human beings before that date or after? It is not a question palaeontology will ever be able to answer. 'The soul' is merely a way of speaking about spiritual awareness, the emergence of the human spirit with a liberated mind, something which will never appear in the fossil record.

The difference between human intellect and animals' may be one of degree, not of kind. Darwin was of the opinion that a 'thinking principle' appears in a number of manifestations throughout the animal kingdom and that there is no radical leap from primate to human being. A slow development in brain growth would lead to the gradual awakening of reasoning powers and conscience.[4] Clearly non-human animals are aware and think, though it is difficult for us to know how to describe the process. It is hard not to be anthropomorphic when speaking about animals and their consciousness because we only know what it is like to be a human being. Some people feel bound to insist that a human mind is in a different category from any other and it is improper ever to speak of mind at all in the case of all other animals even chimpanzees or dolphins. Even then, however, a gradualist account of the development of primate brains can be consonant with belief in the uniqueness of human inner experience. As the chemical structure of the brain increased in complexity generation by generation (perhaps rapidly at some critical juncture of history when a change in the environment made new selective demands on the species) a 'consciousness point' may have been reached. A quantum leap will then have been made from the instinctive consciousness of a primate tied to the present moment, non-reflective and subject to immediate stimulus and response, to the self-aware and more liberated mind of a creature who can ponder on the past and the future, think with imagination and begin to use symbols and language. This 'consciousness point' is analogous to the 'boiling point' of water when liquid is liberated as vapour, or to the 'dew point' when droplets start to form in the saturated air. Gradual change leads at a certain point to sudden change and something new appears. If we are to allow humans to 'have souls', and animals not, then the 'consciousness point' may be the parting of the ways.

Darwin was unable to explain in detail how humans emerged from primates but he did believe that it could be accounted for at least in principle by his theory of natural selection. (Darwin was not an atheist, but his account of evolution did not need God to direct it except perhaps in the very beginning.) He speculated about the liberated effect of upright posture. By becoming bipedal and erect, our hands were left free to become more sensitive and able to manufacture and use tools and weapons. He also proposed an additional mechanism of sexual selection. Along these lines humankind ascended gradually from its primate ancestors.

Darwin's contemporary Alfred Russel Wallace thought otherwise.[5] He was more Darwinian than Darwin in his application of the law of natural selection. This led him, with impeccable logic, to conclude that humankind could not be a product of evolution. God had to have had a hand in it.

As a hyperselectionist, Wallace argued that every feature of a living organism, every feather, digit or behavioural trait has been selected for its usefulness. If he could not account for some characteristic, however odd, then that was due to his ignorance and not to a weakness in the theory. It was a fixation on the concept of usefulness which biologists found hard to shake off in the nineteenth century. For Wallace it amounted to a proof that a divine hand had been involved in human creation because some features could not, in his opinion, be accounted for by natural selection.

Wallace, unlike many of his contemporaries, believed in racial equality. Only culture caused the difference between the European and the savage. But savages who have never known music other than the howling of a war song can be trained to play in military bands, he observed. And the same larynx in a European throat can sing the most beautiful opera. How could such a feature as the larynx have evolved by natural selection thousands of generations before its true usefulness emerged? For the law of natural selection to work it requires that a feature confers an *immediate* benefit on its possessor. The human brain is another feature which in Wallace's opinion pointed to a divine creator. It was created in humans hundreds of thousands (if not millions) of years before mathematicians, philosophers and composers put it to full use. Natural selection, he argued, cannot account for features whose usefulness only emerges in the distant future. Therefore he concluded 'a superior intelligence has guided the development of man in a definite direction, and for a special purpose, just as man guides the development of many animal and vegetable forms'.[6] The God of the gaps is brought into full play as a sort of cosmic dog breeder! It must have been a source of great, if brief, comfort to theology to discover that the theory of evolution by natural selection almost amounted to a proof for the existence of God when it failed at least in Wallace's eyes to provide an account of the human creation. But basing one's belief on *lack* of knowledge has always, as we have already seen, been an unwise policy.

A man is so wonderfully made that Wallace thought it proved God. Some of his contemporaries working in the field of physics, on the other hand, were happily dispensing with the supernatural and with mystery, treating God as an unnecessary hypothesis. Today, one hundred years later, the roles are strangely reversed. While the new physics reveals the deep and mysterious nature of reality and physicists rediscover religious faith, biologists on the whole have gone to the other extreme, reducing people to chemical engines with no inner spiritual life worth discussing. Wallace thought that biology guaranteed humans a unique status but modern behaviourists have downgraded them to no more than one strange organism amongst many others subject to the impersonal laws of stimulus and response. According to this view we are simply a predetermined product of our genes and environment. Our free will, creativity and spiritual sensitivity are aspects of an illusion generated by brain chemistry, and our mind has no more power to guide our actions than a boxer's shadow can influence events in the ring. This dismal apprehension of a person's worth is reinforced by the common knowledge that mental illness, schizophrenia, depression, epilepsy and so forth are all amenable to drug treatment. Personalities can indeed be transformed by pills.

There is a growing sense of dissatisfaction with the behaviourist account of humans. Ultimately we may feel that the term 'epiphenomenal' coined by J. H. Huxley to encapsulate the belief that mind is totally dependent on the brain, with a virtually irrelevant and powerless status, is inadequate. The introspective evidence of mind itself is that it is not merely a non-functional side effect of chemistry following the movements of a mechanical body but never leading them. There is a mystery about the human individual which cannot be described in the language of science (one might as well try to scoop up a river in a sieve). More scientific theory and more research will not get any closer to this mystery because there comes a point when scientific method reaches the end of its road. We already found that this happens when considering the singularity from which the universe is supposed to have emerged. Stephen Hawkins called it the 'principle of ignorance'. It happens again in the complex chemistry of the brain. Neurons and synapses can be described in all their chemical molecular detail (perhaps one day down to the last atom) and the brain's functioning recorded with meticulous precision but the observer will never detect mind. The inwardness of a human individual is not accessible to the scalpel or the electrode. The consciousness of a person remains for ever a mysterious and private phenomenon. Science describes the brain as an electrochemical machine but in so doing only deals with one layer of what is involved in being a person. It is a blinkered view which allows only the language of science to describe reality. Autobiography, poetry, psychology and theology take up the description at other levels and so redress the balance.

One of the most important tasks facing the Christian theologian today is to find a way of speaking about the mystery of a human individual. Neither Wallace nor Darwin were right about the mechanism by which primates became people, though Darwin's speculations came nearer the truth. Wallace gave up the scientific method and concluded that God interrupted evolution to make people out of primates. Darwin on the other hand, convinced that nature does not make leaps,believed that the human species had evolved gradually, undergoing a slow transformation of its stature and its brain size. Today Neo-Darwinism offers a better explanation. Insights into the mechanism of genetic inheritance, into the detailed structure of DNA, the role of mutation and gene shuffling, coupled with a much clearer fossil record of evolution in general gives grounds for much more fruitful speculation. Evolution, it is now argued, does not proceed gradually but in fits and starts interspersed with long periods of changeless equilibrium.[7] There is no need to invoke a long slow process to account for the human emergence, any more than we need to smuggle God in as an explanation. Genetically speaking we are very little different from our chimpanzee cousins (as has been pointed out 99.5% of our genes are the same); only a very small amount of shuffling of the genetic units, or mutation in the DNA of our common primate ancestors, can account for the human branch of our family tree. All morphological change lies at root at this molecular level. A slight rearrangement of the molecular code can lead, it seems, to a dramatic change in the physical form of the descendent creature. It may have been quite a simple matter that created people, such as the postponement of the chemical trigger which shuts off brain growth in the embryo allowing it to become more complex. A few generations may have seen radical changes.

The fact that our brains then remained relatively unused (and no doubt still do) need not lead us to dispense with the theory of natural selection. A quick brain with the power of imagination and the consequent ability to plan ahead must have given our ancestors an instant advantage over their less alert or reflective cousins. In the food and sex stakes anyone with the wit to outflank one's competitors had a high chance of surviving and breeding. The power of imagination may have been one of the significant characteristics which led to the great leap forward from primate to human person.

The full potential of this newly-emerged mental ability would not be realised for millions of years. It had to wait for the development of symbolic language, for the accumulation of a rich vocabulary and for all the meaningful gestures and rituals of a socialising culture. Nature is replete with examples of features which have been selected for one 'purpose' and then turn out later to be useful in an entirely different way; the orchid petal becoming a fly trap, the foot of the mole which became a digging device, the tail of the male bird of paradise originally an aid in flight which is now a sexual signal to the female. Mind, the latest evolutionary emergent quality of chemistry, first appeared (we may presume) by accident conferring a craftiness in hunting or sexual competitiveness on its possessors. The same characteristic after generations of cultural evolution can now be turned to writing symphonies, solving equations or praising God.

We have been considering, so far, the origins of mind, how it emerged in the animal kingdom and whether it is likely to have emerged elsewhere in the universe. But we have not yet asked what mind is in itself. How shall we describe it? What value shall we give it? Although awake to it, or because of it, every day, it is not so easy to analyse. In general discourse we still tend to talk about bodies and minds as though they were two different sorts of substance. Language bewitches one into thinking that the real 'me' – the soul, mind or consciousness –

lives inside my body, which it possesses like a sort of garment. (The phrase 'my body' itself suggests that there is an 'I' to which the body belongs). Descartes the father of western philosophy struggled with and commended this view in the seventeenth century.[8] 'Cartesian dualism', as it then came to be called, was a convenient concept at the time in that it seemed to safeguard a man's moral freedom and kept open the possibility that a bit of him, his soul, might be immortal and go to heaven when he died. Seventeenth-century science was beginning to define humans successfully in terms of the laws of nature. Muscles work like levers and the heart like a pump. The mechanistic image of human beings clanked like a robot over the horizon.

The threat Descartes tried to fend off was the suspicion that if a woman can be described as an automaton, a piece of machinery whose activity is determined by scientific law, then she becomes one more object in a material world. Descartes, by giving philosophical backing to the division between body and soul (mind, consciousness, spirit) came to the rescue. A man's body behaves like a piece of predictable machinery but his soul is free to make moral decisions, free too to float off as a disembodied entity after death. The big problem, however, for Descartes and for other philosophers who championed this dualist view of human nature lay in linking the ethereal soul to the physical body. How can a ghost-like driver pilot the machine? Descartes' own entirely unsatisfactory solution was that the soul influenced the body through the pineal gland in the head!

Today theologians tend to take a different view from Descartes and with many biologists see mind as an emergent quality of chemistry and not as a separate entity. The human spirit, soul, mind and consciousness is an inevitable consequence of evolution. It is an aspect of matter in the way that wetness is an aspect of hydrogen and oxygen when they join chemically as H_2O.

The phenomenon of consciousness leaves science tongue-tied which is probably one reason why behaviourists prefer to disregard it in favour of stimulus-response experiments with the brain. The Cambridge philosopher Wittgenstein put the problem neatly with an analogy. It is as though, he said, consciousness is a beetle which each of us keeps inside a box. Everyone has a box and a beetle but no one is allowed ever to peer inside someone else's box and it is only a convenient assumption that other beetles are like our own. This problem of private access should not deter us. Simply because the tools of science and its principle of objectivity cannot cope with the mystery of inwardness does not mean it should be dismissed as a trivial epiphenomenon.

The alternative to the Christian view offered by contemporary behaviourism is that mind has no great significance and people can be reduced to electrochemistry. This reductionist approach to humanity describes a human being as a chemical engine puffing through a land of illusion on predetermined rails. Only physical processes have any reality and consciousness becomes an irrelevant 'non-functional derivative of physiological change'. The mind is an impotent observer watching the organic machinery run according to its conditioning. It has no freedom of choice, no power to determine events. Finally as the engine fails the conscious mind flickers out of existence deluded to the last that it had some sort of reality.

A sophisticated variation of the 'man is a machine' image has recently become popular with the quest for artificial intelligence, AI. The human brain is an organic computer, according to this analogy. Valuable insights may be gained in two overlapping fields of research as computer engineering and psychology cross-fertilise one another. Psychologists apply the computer model to the complex circuitry of the brain in an effort to comprehend some of its processes while computer experts attempt to create a thinking machine which can follow inductive rather than simply deductive paths of reasoning. The problem for the latter is to design a computer which can make generalisations and learn from its own experience. How can a robot computer imitate the creative, intuitive, hunch-following processes of the human brain? The RAM computer at Brunel University, described in

Reinventing Man by Aleksander and Burnett, is perhaps a first stage in developing an answer to this question.[9] Their computer is linked to a TV camera, analyses and now 'recognises' the faces of members of the engineering department (even if they wear false beards!) The silicon chip revolution, all set to transform society as radically as the agricultural or the industrial revolutions did, raises in a new and fascinating way the mystery of mind in material systems.

So far nothing even remotely resembling consciousness has emerged within a robot computer. Even the most advanced is still almost as far from being human-like as an automatic washing machine and its IQ probably no better than that of an earthworm. A computer's feats of calculation may be prodigious but it is not aware of the process. A programmed robot may outwit a grand master at chess but it cannot even begin to sort the rooks, knights or bishops from the box. But who knows what the future may hold in this field of research? The technology is extremely new and developing rapidly. In a generation or two a self-programming computer with a more elaborate range of sensors and a more complex intelligence system may genuinely come to mimic a human being. Should the computer communicate with its operator using speech – the technology has taken us that far already – then interesting philosophical and psychological problems will arise if the machine claims to be conscious. We are back with the conundrum of Wittgenstein's beetle in a box. How could the operator cross-question the computer to check the veracity of a statement such as 'I know I am awake'? Science fiction fantasies about laboratory-built androids who feel hurt at being treated as second class citizens or computers who reflect on the meaning of life may one day become weird reality.

Will a robot ever be confirmed in St. Paul's Cathedral or will some other world religion be the first to accept a computer as a convert? Far-fetched questions like these may one day have to be taken seriously by theology. Will these creations of technology be 'allowed' to have souls or to claim that they too are made in the image of God? Suppose they do? Consciousness, as we already appreciate, arises from and is based upon complex chemistry. Biological evolution is God's way of putting circuits for intelligence together. The mystery of mind waking up from matter is still, and will always be, one of the richest wonders of the world.

There is no a priori reason why the complexity of a computer robot may not become so sophisticated that it passes the 'consciousness point' discussed earlier. Scientists and technicians will simply have cooperated with the laws of nature and the world as they find it. Such a development may have to wait for some future generation of biocomputers based on carbon and genetically engineered protein rather than on the cold silicon chip. A computer whose AI rises to the level of consciousness will then presumably have to be treated as a person and not merely as a disposable machine. The quest for a new model for comprehending the human brain will then, via the computer image, have brought us full circle to face once more the mystery of personhood.

The Christian faith in this scientific era has to develop its own language to give expression to the belief that we are more than deluded chemistry, that we have a unique value as individuals. Firstly it will explore the significant characteristics of humanity, confident that the deliberations and evaluations of questioning minds, although the products of chemistry, are nevertheless to be taken seriously. Vision is vision whatever its chemical base. Secondly it will find a fine focus in what is revealed in and through Jesus the Jew. We may then find a new framework, true to science, true to human experiences and true to revelation, within which a fulfilling faith may function.

The hallmarks of humanity include self-awareness and the sense of inwardness; the ability to choose one's thoughts and to act with creative novelty; a sense of purpose or intentionality; imagination and compassion; free will and the power to make moral and aesthetic value judgements. These things are not easy to assess or quantify but they are of central importance to any definition of a human being.

In many respects we are not finished creatures: we are

incomplete, our nature open-ended with possibilities for the future. One of the most inadequate aspects of Leibnitz's dictum that this is the 'best of all possible worlds' (so easily parodied by Voltaire) is that it gave the impression of a finished perfect design. This world is a messy place, often chaotic, and therein lies its strength for with the mess there is flexibility and there are opportunities for growth and creation. The image of God is hidden within us but, as the early church father Irenaeus observed, we are not yet God-like.[10] Our true selves are discovered through a disciplined fostering of inwardness as we pursue our vocation to be human.

This vocation to be human involves a great struggle and a willingness to accept grace, as the spiritual literature of the world testifies. A main theme in Plato's philosophy was that education (Latin, *educare* from *ducere*, 'to draw out') involves drawing out, against the resistant forces of ignorance, what lies latently within. In his view, reason should rule over the lower desires of the body and the mental world is ultimately more real and of more value than the physical. In a captivating image he pictured a human being attached as it were by roots from her head to heaven above so that she is drawn upright away from the gravity of ignorance which would drag her down like a brute on all fours to be nearer the earth.[11] St. Paul in the New Testament likewise wrote of the inner struggle between the physical and the spiritual self. The old Adam pulls him back to earth and to death while Christ, the new Adam, raises him up again to share his resurrection. The struggle is not easy, as Paul so ingenuously attests: 'I fail to carry out the things I want to do, and I find myself doing the very things I hate' (Rom 7:15, JB). Tennyson, in the nineteenth century, recognising a man as a product of nature hoped that he might become consciously involved in nature's struggle and so

> Move upward, working out the beast
> And let the ape and tiger die . . .[12]

Early in the twentieth century the Russian orthodox writer Nicholas Berdyaev saw the inner conflict on spiritual terms

'Man is by no means a finished product. Rather he moulds himself in and through his experience of life, through spiritual conflict and through those various trials which his destiny imposes on him. Man is only what God is planning, a projected design . . .'[13]

A human being is not totally definable as a machine because we are part of a process of change. We have a deep sense of new possibilities, that the future is open and that the responsibility for determining the direction of change is settling on our shoulders. The language of contemporary psychology is full of such phrases as 'self-realisation' 'personal growth' or 'self-actualisation'.[14] The obvious appeal of the language and the various therapies that go with it is that they touch on an undoubted longing in the human breast. We are in the state of 'becoming', not yet what we would like to be.

If we are to take part in this struggle with any enthusiasm then we must have a sure and certain faith in the practical reality of our free will. An unswerving recognition of this freedom is one of the most important contributions that Christianity can make to western culture. The mechanistic model has led us too far down the path that defines every thought, word and deed as the mechanical effects of predetermining forces. The behavioural debate about nature or nurture, heredity or environment, forgets a third force – the newly-emerged person who, like God, can act creatively and unpredictably. Through the processes of the universe God is overseeing the birth of colleagues, not manipulated puppets. Unless the self can act as agent then we cannot accept God's invitation to become co-creators.

A sailing boat is subject to the winds and the currents. Left to their own devices these two forces will carry the barque out to sea or perhaps on to the rocks. A sailor, however, in control of the sail and the rudder can use these pressures to her own advantage and, within limits, take the boat where she wants it to go. Emergent mind likewise sits metaphorically above the genetically inherited characteristics of the body and the conditioning effects of the environment. An obvious weakness of this analogy is that it could encourage us to fall back into using dualist language, treating mind as though it were a separate substance. Our new understanding of the 'body-mind problem', however, would suggest that mind is better conceived to be a self-reflective aspect of chemistry. The inner structure of the three pounds of material I call my brain with its 10^{10} nerve cells and 10^{11} glial cells is so complex that a new self-aware phenomenon awakens within it – my mind. No analogy can be adequate to express the novelty of what emerges in the human individual. A new state of being is created. Mind's own self-awareness enables it to transcend the material base from which it has emerged and manipulate its own chemistry and its physical surroundings to its own advantage . . . A man is the most liberated community of molecules on earth; he is emancipated because of his mind. Poetry and theology, scientific research and technology are some of the many consequences of his new-found creative freedom.

An uncompromising enemy of our responsibility for our own spiritual development has been the bad theology of fatalism. It stunts growth and curbs creativity. The conviction that God's omnipotence and omniscience imply that everything had been mapped out from the beginning, predetermined and fixed by decree, received support from the Newtonian picture of the universe. A world composed entirely of atoms which behave like miniature billiard balls subject to predetermining mechanistic laws leaves little room for free will. In such a universe, we are as predictable, at least theoretically, as a machine. But contemporary physics offers us a new image of the atom to contemplate. The quantum picture of atomic particles with its indeterminacy principle suggests that the atom is far from being a fixed predictable entity at all. It becomes, in this view, a diffuse bundle of possibilities any of which may become actual depending on the circumstances and the observer. Every electron, proton, atom, molecule or group of molecules faces not one predetermined future but many alternative futures. The present moment in any material system is open ended in quite a remarkable way. A human being, at one level of description, is a material system and the open endedness of his future is qualified by his creative role as an observer of himself.

There is nothing inevitable about what happens next in this life; a complete definition of the atomic structure of a person's chemistry, body and brain, would give no clue as to which of the infinite number of possible futures becomes real. We have the power, the new physics tells us, to precipitate events in our physical worlds one way or the other. Science which has been the enemy of the doctrine of free will for centuries now becomes one of its best advocates.

The essential character of the world which God is creating is that it contains opportunities to be explored. For billions of years the reconnoitring has been done through evolution by sleepwalkers. The changing forms of plants and animals have explored and occupied the various niches in the environment, without purpose or plan, as unthinkingly as hollows are filled by running water. Evolving life has capitalised on every available opportunity for living space, from the oceans to the skies, from the driest desert to the wettest rainforest. But in humanity the sleepwalkers awake. The human mind is exploring the world of thought at an accelerated rate and is pushing its way into all the opportunities latent in creation. Everything apparently new under the sun, physical or mental, is simply the coming into being of what is logically possible. Mind has produced speech and symphonies, has put men on the moon and released the energy locked in the nucleus of the atom. These and all other cultural and technological advances and developments are merely the uncovering, exploring and refining of what is possible in the unravelling process of creation.

The Christian is bound to meditate on the figure of Christ to understand more deeply the significance of the human spirit. Jesus perceived his own destiny as proclaiming, in the words of Third-Isaiah, 'freedom to captives' (Luke 4:18). He liberates the image of God within us so that like him we may be the children of God, masters not just victims of circumstance (though he shared with us the experience of being victim). One of the most common metaphors in the religious literature of the world is of the sleeper awakening. It is an image Jesus employed in his parables.

The mystery of Jesus' nature perplexed theologians for centuries. Somehow he was a man, vulnerable, emotional, pain-suffering and yet he was also God; two ways of speaking about one person, not two separate natures. It was a paradox. These two aspects of Jesus' character are united by the doctrine of the incarnation, a pivotal concept for all Christian thinking. God's eternal word, his Logos, became a mortal man who walked around Galilee. In Nazareth, his home town, he was disregarded and known simply as the son of Joseph the carpenter; to some of his family he was mad; to his enemies a devil worker and to his followers the first sign of the kingdom of God on earth. In theological language he was 'truly' man or 'very man' and had accepted human mortality to the full by a complete 'kenosis' or self-emptying. His life on earth was not a mere pretence as were the avatars or divine descents in Indian or Greek mythology. The early church rejected the fanciful stories of some gnostic writings in which Jesus wielded strange magical powers, sliding down sunbeams at school or making clay birds fly.[15] Jesus' childhood was normal. Unlike the legendary tales of the Buddha who immediately walked seven symbolic steps at birth, or the God-child Krishna[16] who could open his mouth to his mother to reveal the universe, or even the Koranic account of Jesus' nativity in which the infant speaks from his cot to defend the reputation of his unmarried mother,[17] the real Jesus had to be taught to speak by Mary and to walk at her hand. We may presume that he suffered all the miseries of tooth cutting and dirty diapers, gave Mary and Joseph sleepless nights and generally demanded all the attention any helpless baby needs. Jesus was a mortal Jew like other Jews with (in his case) an ancestry which could be traced back to the family of King David. When he died on the cross on Good Friday at the hands of the Roman occupying forces he really did die.

The anthropic principle revealed by science highlights our significant place in the universe. The laws that govern the cosmos guaranteed that people would appear. If a man in some sense is a key to understanding the universe then the Christian will be keen to know what is revealed in Jesus of Nazareth. It is not blasphemy to suggest that Jesus' ancestry could be traced back to a primate just as surely as he was descended from King David. It is an essential part of his humanity that the genes we share with other creatures he also shares. Two billion years ago an organism which was ancestor of the wheat plant was also ancestor of *Homo sapiens* (we still share recognisably similar enzymes),[18] it was ancestor too of Christ. Whales and mice are our genetic cousins; his also. Jesus was not an apocalyptic science fiction-like superman from the sky but was conceived in a womb and grown from DNA by the laws of chemistry and biology. He was made like any other person in Galilee from the soil and air of Palestine. The mystery revealed by the Christian gospel is that chemistry can be the vehicle of divinity.

Two great movements converge in Jesus Christ, the process of evolution and the life of God. 'From above', as it were, there is the movement of divine creative activity, God's love and redemption. It is a dynamic, flowing process, a loving invasion of the world, enshrined in the iconography of the trinity. The doctrine of the trinity is often misunderstood as though it were a sort of theological snapshot of a divine family (like the Egyptian trio Osiris, Isis and their son Horus, or the three hooded godlets of the Celts, or Muhammad's misinterpretation of the trinitarian formula as comprising God, Mary and Jesus). The doctrine of 'three in one and one in three', Father, Son and Holy Ghost, is a sort of symbolic shorthand indicating a divine act of cosmic significance. We are not intended to conjure up an image of an old man, a young man and a spook! Jesus Christ, the second person of the trinity, is the anointed one (the meaning of the words Messiah and Christ), anointed by the flowing, creating Spirit of God the divine presence which blows in and through the world like a wind.

The second of the movements which converge in Christ comes 'from below'. The rising tide of life, to use the metaphor of Teilhard de Chardin, ascending from the world of matter has produced the human spirit. In Jesus Christ these two great movements join in an extraordinary unity. In him is manifest the meeting point between God and the material universe. We need no longer look to the heavens for a revelation of the divine, for God moves into the world and is found in the life of a man evolved from the earth. Herein lies the uniqueness of Christ and the paradox of his being both God and human which mythological language attempts to express.

A book called *The Myth of God Incarnate*[19] published in 1977, caused an uneasy stir in the church and something of a furore in theological circles. Its media-orientated title caused several anguished commentators to infer that Jesus was being said to be no more than a human being onto whom faith had loaded layers of doctrine. In everyday terms myths are un-truths and so if the incarnation is a myth then it seems there is no way left of talking about Jesus which makes him in principle different from any other person of history. The point that many people seemed either to miss or ignore in this debate is that mythological language in theology is not untrue language but is the appropriate way of communicating insights in matters of faith.

The uniqueness of Christ needs to be restated in twentieth-century terms. This will not necessarily involve the rejection of the mythological language of earlier times but rather its re-evaluation and supplementation. When the first Christian theologians called Jesus 'Son of God', or 'word made flesh', or 'second person of the trinity', they were not using him simply as a convenient peg on which to hang their ideas. He was not merely a humble prophet and healer whom they could cloak with extravagant doctrine. An objective spiritual reality was encountered in Jesus, we may suppose, which demanded mythological treatment. Above all things he was a man who found his true identity in God. That was his experience and not just the way his followers liked to think of him. God is eternal and so Jesus discovered that his true identity also involves eternity. Therefore it can be claimed of him in pauline language that he is the pre-existent Christ, or the eternal word of the fourth gospel; he was 'begotten not made' as in the Nicene Creed or 'uncreated' as in the Athanasian Creed. None of these statements are merely subjective opinions or the results of mind-imposed patterns which make people see castles or dragons in the clouds. They are subjective views of objective truths. Jesus' dual citizenship of heaven and earth seemed, and seems today, a paradox. He is truly a man and truly God. It took the orthodox of the church some four centuries to be clear about the nature of the paradox and to be able to state it with confidence. Today nothing of his uniqueness needs to be discarded in a contemporary account of his person.

According to St. Paul in 1 Corinthians 15:20 Jesus is the 'first fruits' of a cosmic process (cf. Rom 8:23; James 1:18). In him our true destiny is revealed for with his incarnation comes an invitation. That which was revealed in the man Jesus becomes a possibility for all – an extraordinary unity between God the creator and the human individual. The movement of God into the world, symbolised by the doctrine of the trinity, does not stop with the incarnation of the word as Jesus Christ. St. John wrote 'as many as received him, to them gave he power to become the sons of God, even to them that believe on his name: Which were born, not of blood, nor of the will of the flesh, nor of the will of man, but of God' (Jn 1:12–13, AV). In an organic image, suitable perhaps to theology seeking to incorporate the insights of evolution, Jesus is reported as saying 'I am the vine; you are the branches' (Jn 15:5, NIV). This spiritual vine which represents a new step forward of inwardness by the evolving ecosystem will wrap itself around this world just as life and thought have already done. We are, the Christian gospel boldly proclaims, to become co-heirs with Christ. Jesus who shares with us our ancestral roots, built like us from the remarkable chemistry of DNA, invites us to discover with him a deeper reality.

Does Christianity, therefore, have a monopoly of religious truth? Traditionally it has made very strong exclusive claims, and its missionaries have travelled the world with loving though sometimes fanatical zeal. The righteous sense of being the sole possessor of truth even fired the Inquisition into torturing people into the fold of the church. If Jesus is unique, the 'only Son of God', then is Christianity the only true faith and if so, where does that leave the intelligent aliens on other worlds we speculated about earlier? Are they belatedly to be made members of the Anglican communion (which found it so difficult even to admit women to the priesthood) or will they be the recipients of papal encyclicals despatched at the speed of light? I begin to suspect, personally, that truth cannot be such a narrow matter. It surely must contain richness and variety rooted in the recognition that all life, whether on earth or elsewhere, is the product of one universe. There are *no* aliens just as really there are no foreigners.

In John's gospel Jesus is reported to have said, 'I am the way, the truth and the life: no man cometh unto the Father but by me' (Jn 14:6, AV). The statement has often been interpreted to mean that only Christians are acceptable to God. But does it? Jesus speaks throughout that gospel, as its first verses indicate, as 'The word made flesh'. The word is that aspect of God which creates and communicates with the world. It is the living message which was made man in Jesus Christ. But this incarnation of the truth within the Jewish tradition does not mean it is limited to that dramatic encounter with human history. An analogy for it might be the publication of a book. The author's ideas are not necessarily limited to their appearance in print. If they are novel thoughts which will change people's lives they may already be circulating and fertilising the thinking of others. A new creative idea has its own life and the hardbacked volume on the shelf contains just one physical statement of it. Jesus may have been hinting that the work of the word is not limited to its incarnation in Galilee when he spoke in a parable of himself as the Good Shepherd. He then added, 'And other sheep I have, which are not of this fold: them also I must bring, and they shall hear my voice; and there shall be one fold, and one shepherd' (Jn 10:16, AV). It is the universal word of God which speaks here, not just its local epiphany in Palestine.

The Buddhist may hear this word and remain a Buddhist and similarly a Muslim, Hindu or Jew. The atheist too with a deep appreciation of the autonomy of the universe may encounter the word in a way which leads to rejecting all the images of religious culture. A Christian's encounter with the word of God will indeed be through the life of Jesus Christ but that incarnation does not restrict its creative activity.

If we are to suppose that there is intelligent life elsewhere in the universe then the same principle will apply. There is only one word of God in this remarkable cosmos, a universe which has emerged so richly from simplicity. If the universe is an intentional product of the will of God then we should expect that a time will come in the spiritual development of any race of intelligent beings when the word will be revealed in some individual. The word is then made flesh and once more a spiritual vine will begin to entwine its way around the planet where it has emerged. When we *do* contact aliens from other worlds some may already know the gospel story. Nothing will be lost of Christianity's exclusive claims for there is only one God and one truth, but a richer understanding of the inwardness of the material cosmos will have been gained.

Viewed against the antiquity of the cosmos humanity is still in its infancy and once more we should be aware of the newness of the human phenomenon. If we compare the age of the universe to a year then on the same scale the whole of civilisation is accounted for by one second. We have not yet absorbed or accommodated this view point. It means, however, that the gospel on earth is contemporary news, not an old-fashioned religion, out of date and dying. As a species we have only just emerged from the stone age. From flint arrowheads to lunar landings is but a brief episode in the history of life on earth. Satellites to be launched into space by the Indian Space Agency are carried to its cosmodrome near Madras by bullock cart: the leap into the modern world happens while we watch. Only from the standpoint of the new cosmology can we begin to appreciate how new the gospel is on earth. From this perspective Christ has only just come and the good news he proclaims only recently published. The spiritual implications of this event have not yet fully dawned on us. In cosmic terms Christ's day is only now breaking.

We discover that our humanity and new inwardness has a long way to go spiritually. But we do not follow the way on our own. Like the heliotrope flower which has been evolved to follow the sun, we as a receptacle for grace find our true fulfilment in God. The eternal presence, the field in which the laws of evolution have their play has been awaiting this moment since the beginning of time.

Those who follow the way of the word will have to make many conversions within themselves. When John the Baptist invited his contemporaries to repent (Mt 3:2), the word he used meant to 'turn around', and later St. Paul instructed new converts to 'be transformed by the renewal of your minds' (Rom 12:2, RSV). Our power of imagination and our faculty of reason must both be dedicated to the word if we are to follow the way. The mind, that feature which sets us apart from other animals and gave us in the early days great selfish survival value, can now be dedicated to a new task, the love of God and the service of other people. The power of imagination, which enables a creature to transcend time and space, put humans well ahead of competitors in the quest for sex, food and security. Guided by the word this power can be transformed and put to other uses and higher purposes. It is only with imagination that we can begin to transcend our own selfish standpoints and love our neighbours. Imagination makes it possible to leap over our solipsist walls and enter other people's worlds.

In our new-found freedom we are creators of values. There is a rational component in ethics and so our innate love and altruism will be directed by reason and not merely by the unconscious pressures of a behavioural trait which has helped us survive as social, group-orientated animals.[20] Jesus Christ who proclaims freedom to captives can liberate our minds. Freedom to find ourselves in loving others; freedom from anxiety; freedom to be born again and grow, to heal and be healed, to forgive and to be forgiven. With this new liberty comes the guidance of the creating Spirit who sustained us long before we knew it. And with the Spirit we gain a new authority to explore and interpret, as Jesus himself did, our own religious traditions to find in them what is good and true. Jesus Christ who evolved from the earth *and* descended from heaven – is a guide in this autonomous material universe. Inwardness is an emerging spiritual fact of the universe and mind reveals a longing for a deeper reality which only the word made flesh can satisfy. People are not just another sort of thing. The highest value and the greatest source of meaning in the known universe is to be found in humans. The dynamic symbol of the incarnation in Christian theology nails our attention right there.

6. CATASTROPHES AND CANCERS

'You can't have good without evil', it is often remarked. It is rather as though evil were analogous to an artist's use of shadow in a picture, to help reveal form and give substance to the highlights. A happy woman may be satisfied with this answer to the problem of why a loving and all-powerful God allows horrible things to happen when God could stop them. But it wears a bit thin if the evil is experienced directly, and even the composure of a happy man may be shaken from time to time and his blood made to boil with anger at injustice in the world. Mindless cruelty is reported daily by the press. On 26 February, 1985 *The Guardian* carried this item:

> A Belgian couple went on trial in Mons yesterday charged with murdering a nine-year-old Zairean boy in their care after subjecting him to six months of torture. Francis Baudoux, aged 38, repeatedly kicked and beat the boy in the face, dropped him into a bath of boiling water and forced him to eat his own excrement, the court was told.

(They were given life sentences.) Agony, particularly the agony of a child, raises the question 'Where is God?' with powerful insistence.

Suffering is intrinsic to much of human experience and nature is full of disaster and pain. A scientific attitude to people and their environment offers the possibility of a new approach to what has been called the problem of evil. Anyone who believes that the universe should be interpreted in exclusively materialistic terms and who thinks that a scientific description has the last word may have no problem with suffering, at least intellectually. If life is no more than an unintended chemical oddity and ultimately meaningless then painful accidents that happen on the way are just too bad. They require no justification. It is only with a positive spiritual faith, believing in the reality of God, that suffering becomes a problem. If God is good and is responsible for the evolving universe then why does God allow cruel tragedies such a prominent place on the stage? Various tentative answers have been advanced. In the case of Christianity the focus has been upon the enigmatic figure of the suffering servant, the Christ who is crucified. We who now live in a scientific era have to answer the question once more in our own way and from our own perspective. We will have to consider all sorts of evil, not only human cruelty but the great catastrophes of nature and illnesses such as cancer. What we say about God will have to be consistent with what we know about the world, and science will have a unique contribution to make as faith rationally explores its religious traditions and beliefs.

If we are right to believe that we are morally free agents able to make genuine choices and value judgements, then one of the inevitable and frightening consequences is that we will have to accept too that God will allow us to experience the effects of those choices. The sadistic torturing of children and the gas chambers of Auschwitz are among the ghastly results. Terrible as they are, however, they can be seen to be the logical outcome of the creation of a world in which there are morally free agents. The responsibility lies with us to see that such horrors do not happen again. They are a challenge to our integrity and our humanity, a goad to our love.

Harder to justify are those experiences of pain and suffering which are not consequences of our free choice but are due to aspects of the way God has created and is creating the world. Could God not have made this world more disaster-free, less painful, less 'red in tooth and claw', less prone to catastrophes?

The garden of Eden story in Genesis suggests that pain, toil and death entered the world because of humans disobedience. The perfect creation willed by God was spoilt when man and woman ate the forbidden fruit. Death was punishment. Charles Darwin with his theory of evolution by natural selection gives us an alternative and very different view. Death according to this account is a necessary and positive part of the process of change. Without death and the replacement of one generation by another there could be no development by natural selection, and evolution would never have produced people. It must therefore be part of God's plan that death should have a creative role to play as God wills our changing world into existence.

Megadeath has recently begun to fascinate people today. The extinction of a whole species is not limited to the dodo, that fat flightless pigeon which was so easily and unintentionally wiped out in Mauritius in the seventeenth century. Darwin after considering the phenomenon of the demise of the giant quadrupeds of North America, concluded 'certainly no fact in the long history of the world is so startling as the wide and repeated exterminations of its inhabitants'.[1] More than ninety-nine percent of all the species of creatures which have inhabited the earth no longer exist.

In the 300 years since the last dodo was sighted in 1680 there have been at least 300 extinctions of other vertebrate animals, more than half of these being full species. The pace of megadeath is increasing rapidly as humanity asserts its presence across the surface of the planet. Estimates of the number of extinctions among plants and invertebrates by the end of this century vary between 450,000 and 1,800,000.[2]

When will it be humankind's turn to become extinct? Will we manage to avoid this fate? Dinosaurs roamed the earth for 120 million years before vanishing from the scene. We have only been around for two to three million years. Will our intelligence enable us to do better than the dinosaurs or will we overstep ourselves, our mental agility and inventiveness being the source of our downfall? The threat of an exterminating nuclear winter following a nuclear holocaust has come to fascinate the media as we approach the end of the twentieth century.

Humankind may die at its own hands. If it be by global

nuclear war then we will have misused our freedom (a risk God has to take) and deprived countless future generations of the opportunity of health and happiness and of discovering the kingdom of heaven. This era we live in may be the most danger-ous in the history of the human race. If we manage by a mixture of luck, good sense and prayer to survive it, then I believe humankind will have a billion years or more of exploration, creation and discovery spreading out ahead of it. In the words of Lewis Thomas, the optimistic biology watcher, we may look forward to ' . . . a whole eternity of astonishment stretching out ahead of us'.[3]

But a nuclear holocaust is not the only way humankind may meet its doom. Disasters and accidents on a grand scale are regular features of God's world. People used to think that great calamities were due to the influence of the stars, hence the word 'disaster' (Greek, *astron* for star). Today we see them as consequences of the regular laws of nature.

Orbiting silently in the dark wings of the sky, a catastrophe may await Planet Earth. Like the slow circling of a buzzard it swings in great arcs around the sun waiting to stoop and kill. A mountain-sized asteroid's flight through space is so smooth it would seem motionless if you were to stand on it. Dust and rocks hold lightly to its surface under the tug of a gentle gravity. It may be coated with frost or be as black as soot as it tumbles slowly beneath the stars. Not a whisper of wind disturbs its silent, airless sky. The blazing sun rises and sets every few hours on its jagged edges, its dark valleys and cratered highlands.

Against the backdrop of a velvet black, star-studded sky the earth will one day loom large – a perfect blue and white ball of colour and brightness. Only as the asteroid plunges into the atmosphere will the dust get swept in a hurricane from its surface; the rocks will heat and glow with fire and the roar of a million thunder storms will echo round the world. The impact will dig a crater miles deep and if it should be in the sea the salty waves will cascade into temporary orbit; thousands of cubic miles of powdered rock and pebbles will fly up from the impact around the world. Some larger rocks may reach escape velocity and fly to the moon. A high haze of pulverised rock dust swept by winds from horizon to horizon will cool the heat of the midday sun. Hundreds of billions of tons of this rock dust as fine as talcum powder will encircle the globe high in the stratosphere. Vivid red sunsets will dominate the evening skies for years; not the lurid angry red on the undersides of dark clouds due to moisture in the atmosphere but a deep redness right across the sky above and beyond the clouds. If there were any shepherds left to take delight in it they might mistakenly conclude that the red sky was a sign of good weather to come.

In low countries and by sea margins there will be devastation on all sides, the landscape a wild jumble of wreckage and flotsam, the havoc wrought by tidal waves hundreds of metres high that thundered in from the ocean and pounded the land. After such an event the world would never be the same again. There would be mass extinctions of creatures and plants and it would take years for the world's climate to settle down and return to normal. Humanity might not survive such a catastrophe.

Goya's painting of the God Chronos, Time, devouring his children like a ravenous giant, blood dripping from his mouth, comes more easily to mind as a picture of the creator of such a catastrophic world than does the loving Father, of Christian teaching, who provides for his children. Cosmic vandalism, death and destruction are hallmarks of the process of creation. The question of an asteroid's impact with Planet Earth is not merely academic. Such disastrous events have happened a great many times in the past and are quite certain to happen again – though much less often. It is speculated on good grounds that the age of the dinosaurs came to an abrupt end because of an asteroid collision, or if not an asteroid then a comet which amounts to much the same thing. A comet, which may be described as a giant dirty snowball, is a conglomeration of ice and rock about the size of a city. The comet's traditional tail is simply a faint streamer of gas and dust blown away from the comet's head by the solar wind which fans out gently from the sun. Like an asteroid, a multibillion ton comet would crash to earth at about thirty or forty kilometres per second.

The dinosaurs died out about sixty-five million years ago and their demise coincides with the boundary between the Cretaceous and the Tertiary, two great epochs of the earth's geological and ecological history. The saurian reptiles were not the only creatures to be hard hit in this great extinction, for there is evidence of mass exterminations on a global scale. Something like seventy percent of all living forms vanish at this point from the fossil record and new ones take their places. Geologists collecting and analysing samples of sedimentary rock of that distant era have uncovered unambiguous evidence of a significant boundary event. The same clear story emerges all over the world. A distinct line of rock separating one layer from another marks the boundary between the Cretaceous and the Tertiary. It is exposed, for example, for all to see, in the low cliffs of the seashore at Studland in Dorset. Analysis of the compact clay that comprises the boundary layer reveals that it is enriched with traces of two noble metals iridium and osmium. This evidence has been interpreted to mean that the layer contains the disintegrated remains of a devastating visitor from space, an asteroid or comet. It caused the sun to set on the Cretaceous era and cleared the site for the evolution of the mammals.

Dozens of theories have been advanced for the great dinosaur extinction; a supernova stripping the earth's atmosphere of its ozone layer exposing the surface to the frying and cancer-producing ultraviolet radiation of the sun or a reversal of the earth's magnetic field with similar results; a change in climate due to an interstellar dust cloud cutting down the sunlight or, more surreptitiously, a plague of tiny night-roving mammals which ate the dinosaur eggs. The theories themselves would fill an intriguing and informative encyclopaedia. It is the theory of the earth's collision with an asteroid or comet, however, which carries the strongest credentials. Not only is the evidence from the boundary rock layer convincing but what we know of the formation of the solar system leads us to expect catastrophes and exterminations on this vast scale. We only need to look at the moon through binoculars to see some of the ancestral scars caused by asteroids.

Asteroids are some of the inevitable pieces left over from the days of planet formation. They are the building site litter of the solar system, an unavoidable consequence of gravity's creative work in constructing suns and planets from immense clouds of interstellar gas and dust. Five billion years ago this gas and dust formed rings round the newly burning sun and within this orbiting material millions of lumps began to take shape. They were the proto-planets known as planetesimals. In time the largest of these conglomerates of matter, formed by gravity, attracted the surrounding debris to themselves and grew in size. In this way the major planets were built up. Many pieces left over continued to orbit the sun on their own, colliding occasionally with one another. These are the asteroids. There are thousands in the solar system, ranging in size from those big enough to be called minor planets such as Ceres a thousand kilometres across or Vesta at half that size, to lumps of rock the size of a house or a kitchen table. The majority orbit the sun between the planets Mars and Jupiter and are probably the debris of larger objects which after many collisions have broken into irregular-shaped boulders. Eros, for example, which I watched through a telescope one winter tumbling end over end through the constellation of Gemini, is a mountain range thirty two kilometres long by only seven kilometres wide.

The moon and the planets Mercury, Venus and Mars, and also the satellites of Jupiter and Saturn, are still heavily pock-marked from the multiple impacts of these flying mountains at an early epoch in the history of the solar system. The face of the 'Man in the Moon' is formed from the dark scars of some of the most devastating of these collisions. The only reason our own planet, earth, shows few signs of such cratering is because the erosion of weather and the slow movements of the earth's crust have wiped out almost all the evidence.

The great bulk of asteroids lie beyond Mars, but there are a few which orbit closer to the sun than that. The group whose annual trajectories take them across the earth's orbit and nearer to the sun than us, have been named the Apollo asteroids. Icarus is a typical example; the size of Mount Snowdon, it orbits the sun in three and a half years at something like thirty kilometres per second. In 1970 it passed within only a few million kilometres of earth. Other closer encounters lie ahead of us.

In the year 2015 an asteroid-like object (it may be the rocky core of an old comet) will pass between the moon's orbit and the earth.[4] Named 1983 TB it has the same trajectory around the sun as the clouds of tiny particles of debris which give rise to the Geminid meteors when the earth encounters them annually in early December. The small grains of rock which burn up in our atmosphere appear to us as shooting stars and all seem to fly across the sky from the constellation of Gemini; hence their name. Someday, if not in the year 2015 then in the distant future, 1983 TB may conceivably collide with our planet and release more energy than would be let loose by a global nuclear war. The gentle Geminid meteors will have heralded a storm as sparks before a forest fire.

Asteroids and comets sometimes change course dramatically. The disruptive influence of a possible tenth planet, Planet X (or should we call it Thor?) could cause comets from the postulated Oort Cloud, a field of millions of comets suspected to be hovering out beyond Pluto, to plunge into the inner reaches of the solar system.[5] If not Planet X, then Jupiter might suddenly hurl an asteroid towards earth, the gravitational field of the massive planet swinging the asteroid into a new trajectory like a slingshot. (NASA scientists have used this mechanism to redirect the Voyager spacecraft *out* of the solar system). It would be an irony if humankind, after avoiding a nuclear holocaust, were finally wiped out accidentally by the celestial missile of a pagan god.

But would God let such a catastrophe happen today, one that could exterminate the human race which is still a relative newcomer to the planet? The more we learn about this universe the more we discover what a perilous place it is. Eighteenth-century theologians looked for the beauty and harmony, the machine-like regularity and all the providential adaptations in nature which seemed to shout loudly that a designer, a cosmic architect, was behind the universe. Nineteenth- and twentieth-century theologians have been more realistic. God is not so easily exposed. The universe is a hard, often horrible, place and indiscriminate in its catastrophes. Earthquake, flood and plague are no respecters of persons. The rain falls on the good and the bad alike; even giant rocks rain from the skies.

Disasters and purposeless destruction are an unavoidable part of the way of the world. I raise the issue of a wayward asteroid or comet because the devastation created would be equivalent to that of a nuclear holocaust. Yet it would be totally undeserved and bear no relation to the way we have conducted our lives on earth. We could not attribute it to our misuse of our free will; ultimately if anyone is to blame it has to be God.

Let the asteroid collision be a parable of all the calamities that we read about in the daily press. Catastrophes hit the world and individuals with unremitting frequency (and at root all disasters are suffered by individuals). Some are caused by people, others by nature. The tortured child suffers because of the warped psychology of his inhuman guardians. By 1982 ten million people had been killed in the 140 wars that have scarred the planet since the end of World War II. In 1971 a million people, men, women and children were drowned when a cyclonic trough of low pressure led to a storm surge swamping the delta of the river Ganges in Bangladesh. By 1984 fifteen million people throughout the world had died as a result of car accidents. Millions suffer eye disease and blindness in Africa from bacteria carried in water. Hundreds of thousands starved to death in the 1970s in the Sahel, south of the Sahara, when the monsoon failed to bring rain from the Atlantic and drought parched the land. Similar tragedy hit the people in Ethiopia a decade later.

Whether buried beneath the rubble of a bombed Beirut, or lying under the demolished shambles of a town shaken to bits by an earthquake, the nameless statistics become a regular part of the television news. The disasters are shocking but they only become harrowing when we can put a name or a face to the tragedy. In 1979 an Italian child, Alfredo who fell a hundred feet down a narrow well and got stuck, tugged at the hearts of millions. His family and those attempting to rescue him could speak with him and hear his tears for days over a radio let down on a rope. But he never came up alive. A thousand people, nameless to us in England, died in Iran in an earthquake at the same time. They remained for many an empty statistic compared to the death of that one little boy. The fact is we harden ourselves easily against the many messages that come to us from a cruel world. In T. S. Eliot's words 'humankind cannot bear very much reality'.[6]

Would God allow a mountain-sized asteroid to devastate the world? The answer is that God already *has,* many times, and permitted much else that is like it. And there is no reason to suppose that just because there are now people on earth God will prevent any future accidents. It would certainly be comforting to think that God could, and would, give the asteroid a hefty celestial nudge (after due consideration of the number of righteous people currently upon earth) and send it hurtling into the sun. Would church people sing litanies expecting a providential divine intervention? 'Lord send us seasonal showers but not the destroying rock.' Perhaps a world day of prayer would be proclaimed focusing from St. Paul's Cathedral onto the sky. The re-routing of such a cosmic juggernaut would be a dramatic demonstration of divine presence and power. But it is only in legend that God stopped the sun in the sky to give the Israelites time to slaughter their enemies (Josh 10:13) and split the moon in half to demonstrate the prophethood of Muhammad. In the real world dramatic signs like this are just the sort of thing which God, according to Jesus, refuses to give.

An enlightened approach to the painful reality of undeserved disasters began with Job. He was a good man brought up to believe that God rewards righteousness and punishes the wrongdoer; an expectation spelt out clearly in the book of Deuteronomy. Yet Job was beset by calamity after calamity (Job 1:1–19). Unprovoked aggression by raiders, lightning bolts and tornado; one after another they destroyed his property and killed his children. Finally he was afflicted with malignant ulcers all over his body and with a wife and friends whose attempts at comfort were so inadequate they only made matters worse. The capricious cruelty of circumstances would have made many lose faith.

This world is not always the safe providential place that a simplistic belief in God might suggest. The story of Job is taken up by Christ who suffered undeserved pain and rejection; the cross, the symbol of his passion, borne willingly, became a pivotal motif in the Christian gospel. In the collection of Jesus' sayings known as the sermon on the mount there is the suggestion that good and bad fortune are indiscriminate, 'he causes his sun to rise on bad men as well as good, and his rain to fall on honest and dishonest men alike' (Mk 5:45, JB). Later, not long before he died, he warned his disciples that the world would not be an easy place to live in: 'When you hear of wars and rumours of wars, do not be alarmed, this is something that must happen, but the end will not be yet. For nation will fight against nation, and kingdom against kingdom. There will be earthquakes here and there; there will be famines. This is the beginning of the birthpangs' (Mk 13: 7–8, JB). Here is no soft option or escapist dream.

We may be waking up to some of the lurking dangers that threaten our existence only just in time. There is no protective guarantee built into our genes that gives us immunity from extinction unless it be the wise use of our rationality. We are the first creatures able to exert conscious control over our environment, to be masters and not merely victims of circumstance. If it happens one day that a multibillion ton rock is found to be on a collision course with earth it will present humankind with a gripping problem. We are the first generation of people who might be able to do something about it. We have begun to conquer space, the high frontier, and we have learnt how to harness the immense energy locked within the atom. A space mission to intercept the asteroid, using the combined technological expertise of the super-powers, could either blow it up using nuclear bombs or else deflect it into earth orbit. (The joint venture against a common enemy might even bring peace to the world!) As a tiny new moon in orbit the asteroid could be of incalculable value. Some asteroids are composed predominantly of iron and such a one would prove a timely harvest from space solving the world's iron needs for hundreds of years. The point is that without the science and technology which we are just now developing (in nuclear research and space exploration) we would only be able to sit back and wait for catastrophe to strike. Whatever the current threat posed by the invention of nuclear explosives and defence systems, if we manage to avoid their misuse the capability we have developed may in the long run be invaluable.

Natural disasters have had an interesting history in theology. The archetypal catastrophe is the flood, that act of God which destroys life, property and communities. Flood stories are found in almost every ancient culture because flooding from a variety of causes is a widespread phenomenon. There is absolutely no need to suppose, as biblical fundamentalists do, that these stories all hark back to a single worldwide flood from which only Noah and his family escaped with their lives (Gen 6:9–7:23). Noah is one of a common type. He and his family were the lonely survivors of what, to them, was a great catastrophe, and survivors always have a tale to tell. The universal fascination with death and disaster guarantees them an audience. The various legends and myths that come down to us from the past (Indian, Sumerian, Hebrew, Greek, Native North American and so on) come from eras when communities of people lived in their own small worlds. A river flooding from horizon to horizon across a lowland plain would swamp the whole world as far as the sedentary village-dweller was concerned. The same is true for a storm surge rushing inland over low-lying coastal plains. Amongst the Californian Indians the break up of a temporary dam caused by a retreating glacier high up in the Rocky mountains would send a wall of water down the valley, washing away villages and trees as it went. There may even be in some of the legends echoes of earlier stories from the time, ten thousand years ago, when the ice cap covering Europe and North America melted and retreated.

Behind many flood stories there often lies a deeper question about why the God or gods sent it or allowed it to happen. (This theme is not always central and sometimes other details will take over. So, for instance, the Smith River Indians of Northern California have a flood story where the emphasis lies on how the two surviving Indians coped with a world devoid of fire. No longer able to cook their fish, they resorted to warming them up by carrying them around beneath their armpits.)[7] A dramatic catastrophe out of the blue frustrates people's search for meaning in life and poses the question 'Why?' It is hard to stomach that a world-shattering event has no reason; haphazard and arbitrary disasters like floods are difficult to bear. Many flood stories, then, represent humankind's attempt to cope with catastrophe, and the question 'Why?' is one that people continue to ask today long after the times when the answers provided by the first flood stories have been rejected.

One of the earliest flood stories to ask this question is recorded in cuneiform writing on clay tablets found in the great library at Nineveh, where they had been collected in the seventh century BC by King Ashurbanipal. Other fragments of the same tale are even older and date back as early as the second millennium BC. They have turned up in many places from Turkey, Palestine and Mesopotamia, and in a variety of local languages, Indo-European, Hittite, Semitic and Akkadian. This story of the flood is one chapter in a longer tale called the Epic of Gilgamesh, an account of the wandering of a legendary hero who was King of Uruk. In his journeying he searched for a man whom the gods had nicknamed the 'Faraway' because they had sent him to live in the distance at the mouth of the rivers. This man, otherwise called Utnapishtim, with his family and some animals, had survived the great flood in a boat which he had built after being warned of the deluge to come. In its details the story that Utnapishtim tells is remarkably like the story of Noah in Genesis. A favoured man, warned of the deluge about to break on the world, builds a boat, packs it with family and animals, survives the rains and the storm, lands finally on a mountain, sends out birds (even a dove) to see if the waters have subsided anywhere, and then on the top of the mountain makes a thanksgiving sacrifice for his survival. The rainbow in the sky, God's promise to Noah that he would never do this again, is paralleled by the Queen of Heaven, Ishtar, holding her great sky necklace, and promising by it that the gods would never forget the disaster that they had wrought upon the human race.

It was once thought that the parallels between these two stories were simply due to the fact that the Hebrew account was a retelling of the older Sumerian tale. It is more likely, however, that there were a great number of flood stories around the Mesopotamian area, where flooding was common. Archaeological digs reveal plenty of evidence of such extensive inundations many millennia ago. The two accounts we have simply follow the conventional pattern of storytelling. The details that differ depend upon the culture and beliefs of the storyteller. So the Hebrews have one God who sends the flood whereas the Sumerians attribute it to the decision of a council of gods. Utnapishtim takes on board animals both wild and tame, while the Hebrew tellers are more concerned with the kosher dietary laws which had by then become part of their social pattern; Noah takes on board seven pairs of clean animals but only one pair of unclean animals. The most significant difference in the story however lies in its answer to the question 'Why?' In the Sumerian tale there is no suggestion of morality or justice; the gods decide to wipe out everyone because they have become a noisy inconvenience downstairs, rather like unacceptable tenants. 'In those days the world teemed. The people multiplied. The world bellowed like a wild bull and the great God was aroused by the clamour. Enlil heard the clamour, and he said to the gods in council; "The uproar of mankind is intolerable, and sleep is no longer possible by reason of the babel." So the gods agreed to exterminate mankind.'[8] The gods in this tale are no more than inflated versions of selfish human beings motivated by self-seeking emotion and bad temper; like inconsiderate landlords they evict their inconvenient and unruly tenants.

The story of Noah indicates a development in moral thinking. The authors of the first five books of the Bible, the Pentateuch, were convinced of the oneness of God, that God ruled the world, whose purposes were totally just. Their answer to the question, 'Why are there floods that kill people?' is cast, consequently, in moral terms. The people on earth were not destroyed because they were a noisy pest but because they were wicked and not walking in God's way. Only the righteous man Noah was allowed, with his family, to survive the destruction of the polluted world (Utnapishtim survived not because he was good but because he was fortunate enough to have a household God, Ea, who tipped him off in a dream about the coming deluge). Noah with his ark load of family and animals is then the means for repopulating a newly cleansed post-diluvian world.

Ideas about God and the world change and evolve like anything else. Floods, we now see, are a consequence of the laws of nature – the purposeless results of blind chance – and not due to supernatural intervention. No one today, with any sense or sensitivity, suggests that people who lose their lives in floods are being punished for their wickedness or the hardness of their hearts. The question has swung from 'Why the flood?' to 'How did it happen?' We have a much greater understanding of the weather machine today than did our ancestors and we are able with some success to predict the cyclones and tornadoes of tomorrow and to give an even better explanation with hindsight. Computers, satellites and an internationally coordinated system of weather stations enable us to have a global view of our climate and its day to day activity. According to the way we now look at the world, disasters such as floods or storms which destroy property or kill people, follow predictably, at least in principle, from the known laws of nature. A slow process of waking up to reality has been gathering momentum, and with it the possibility of a new spiritual maturity.

News of catastrophes in foreign lands or from past history are not the calamities which call our faith in question most strongly. It is the personal tragedy striking out of the blue that is the most disturbing; the cancer that kills a young mother for instance. Cancer is a subject which our open permissive society finds difficult to discuss, despite the fact that almost everyone has a relative who has died of it. Even the name of the disease has a malignant potency for fear. Asteroids may be a by-product of planet building, floods a consequence of the climate, and so inevitable hazards, but surely God could have designed a world in which cancer did not exist?

At first sight cancer might seem to be a disease of civilisation and therefore more our fault than God's. A society which sent little boys up chimneys as sweeps, a disgusting job on all counts, gave them cancer of the scrotum from the carcinogenic effects of soot. Strontium 90 from nuclear bombs invades the body by replacing atoms of calcium in the bones and then kills from the inside as it radiates. Cigarette smoke precipitates cancer of the lung. Asbestos used in housing and the brake linings of vehicles powders into a fine haze of tiny hooks which, getting into the lung, can cause mesothelioma with disability and death years later. Even wives who have never entered an asbestos factory have contracted the disease through washing their husbands' overalls, a hard reward for performing a loving task. Statistically, a greater percentage of people die of cancer in the developed industrialised part of the world than they do in third world countries. As usual, the bare statistics mislead. In developed countries two other facts have to be considered: many other illnesses that killed people in the past have been eliminated or effectively controlled; and in countries where there are more doctors the number of deaths diagnosed as being caused by cancer will be significantly increased. Cancer cannot be dismissed as an evil of civilisation and due simply to the mismanagement of oneself and the environment, for, like the wayward asteroid, it is a natural part of the world as we find it.

All organisms built from cells can get cancer, whether oak trees or oysters, fishes or birds; the disease is simply the malfunctioning of the cells from which living things are made. Cancer is not a disease like measles, malaria or smallpox which can be eliminated by a coordinated plan of treatment (the last smallpox case being recorded by the World Health Organisation in 1980). Unlike most other illnesses there is not one simple cause of cancer and contemporary research reveals a multitude of ways that it can begin. For 1,300 years the medical profession was ruled by the opinion of Galen, who in the second century AD claimed that health depended on the right balance within the body of the four humours. He speculated that an excess of black bile was the cause of cancer. No-one bothered to research the matter any further because it made a sort of neat sense, and so Galen's judgement was not really contested until the Renaissance. More recently bacteria were thought to be the cause of cancer, and then viruses. Today we know that the causes are much more complex and cancer is described as an accidental development within the reproductive process of cells that may be the result of any number of causes, in some cases many working together. Virus infections, diet, genetic tendencies, toxic chemicals, irritants, even psychological stress can all separately or together be triggers of different types of cancer in different parts of the body. Just as there is no one type of event which we call an accident, so there is not one kind of cause and effect which is cancer. The word covers a whole family of diseases, all to do with the malfunctioning of cells.

Cancer is a phenomenon of faulty reproduction and can only develop where new cells are being manufactured. Parts of living bodies have to be continually replaced, otherwise they would die quickly. The bone marrow is manufacturing blood cells all the time to replace those that are battered to pieces as they churn around the arteries and veins of our bodies; their life span is only a few months. The intestines have a pretty tough time processing the food we push through them. Our skin needs regular replacement as it gets abraded away in a rough world. So it is in these areas where new cells are being manufactured in a continual hive of activity that sometimes the complex processes can go wrong and cancerous cells grow.

Cancer may fill us with disgust and anger particularly when it afflicts someone we love. This makes it even more important, if we want to understand the world we live in, to see it in perspective and eventually to discover that it is possible to contemplate cancer with some equanimity. It is only against a background of a deep appreciation of the mystery of life manifested through the organisation of cells that we can cut cancer down to size. Recent research has revealed some of the incredibly complex and beautiful processes that take place within a cell as it relates interdependently to its neighbours. A cell is a turmoil of activity, an unresting busy process, and a still photograph tells you no more about it than a camera on board an aeroplane can record the life of a city.

One of the most remarkable parts of the cell is its nucleus which appears in a still photograph as an amorphous hard grainy lump. It contains the molecules called DNA discovered by Francis Crick and James Watson in 1953 to be in the form of a double helix. The DNA is bundled in forty-six packages known as chromosomes. In simple coded form these molecules contain all the information needed to make a whole human being. Height, colour of skin, number of legs, shape of nose, type of hair, structure of brain, even the age at which you will tend to go bald, is all stored in a more efficient data system than has yet been devised in any computer. Only four simple chemical bases are needed for spelling out the code from which all the information can be read. If the same information contained in the DNA of one human cell were to be written out in English it would fill, it has been calculated, 1,700 volumes of the *Encyclopaedia Britannica* or 10,000 miles of computer print-out.

The nucleus floats in what seems an inchoate watery soup,

the cytoplasm, but which in reality is an intricate and highly organised system of fibrous protein molecules. Cucumber shaped mitochondria wriggle their way around within the cytoplasm as almost independent mini-monsters and miracles of micro-biochemical engineering. They transform sugar into energy while RNA molecules, messengers from the DNA in the nucleus, synthesise protein. Each cell is a city of activity with its own data banks and factories, and yet we are producing them in our bodies at millions per minute. The greatest renewal is in the skin, in the intestines and in the bone marrow. At every cellular division (except in the case of red blood cells which have no nucleus) the coded information in the DNA has to be reproduced accurately without misquotation or spelling mistake. The double helix is unzipped so that each half then forms the basis for the new complete DNA of the daughter cell. One of the most remarkable and so far not entirely understood mysteries is how each cell knows where it is in the body. The nucleus of every cell contains the same DNA, yet the cell 'knows' somehow that it has to reproduce as a cell that is part of an eye or a toe, a gut or a muscle, a bit of brain or a bit of brawn. Out of all those encyclopaedias of information, how does it know which page to read? If it were not for this process of differentiation whereby cells function in very particular and precise ways in different parts of the body, cell division would lead to no more than an amorphous and amoeba-like lump of shapeless jelly. Life would have evolved no further on earth than as some formless slime. Morphogenesis, or what gives the organised shape to living things, has become an intriguing area for speculation and research. The answer hangs to some extent on what is happening in the membrane which envelops the cell. This membrane is not just a plastic bag-like skin designed to hold everything together and stop the cell from falling apart. It is through the membrane that the cell communicates with its neighbours, importing and exporting goods and discovering where it is in the body and therefore what part it is expected to play in the whole organism (through the use of hormones, perhaps, or local electric fields). What a cell does depends as much on where it is and what its neighbours are doing as on any intrinsic property of the cell itself.

It is against the background of these mysterious processes of life and the wonderfully intricate nature of the cell and its ability to divide and reproduce, that the problem of cancer must be seen. Given the phenomenal rate of reproduction and the incredible intricacy of the internal organisation of a cell, it is perhaps remarkable that malfunctions do not happen more often.

It is when the cell's relationship to its neighbours breaks down, so that it is no longer regulated or inhibited by them, that it becomes isolated and we tend to see the development of cancer. How this happens, and whether the isolation is a cause or a consequence, is one of the subjects of contemporary cancer research. What is quite clear, however, is that there is no one simple cause, and no one single type of cancer. The word cancer is simply the term used of cells which, having become isolated from their neighbours, begin to reproduce incorrectly. From their own point of view, one could almost say, there is nothing wrong with them, any more than there is anything wrong with flowers that are labelled weeds in a garden. They just happen to be blooming and seeding themselves in the wrong places. The cells' behaviour from then on is lawless and their subsequent divisions and growth are no longer related to the needs of the organism. They may spread to other parts of the body, migrating via the blood stream or lymph channels, and form equally anarchic secondary deposits known as metastases. The causes of this lawless behaviour amongst cells are as multifarious as are the accidents that happen to living creatures of a human dimension. A great many have been reliably traced: radiation, spontaneous errors in genetic reproduction when the cell divides, the presence of carcinogenic (cancer-producing) substances such as asbestos or some, (now banned) insecticides, viral infections, viruses smuggled into genes rather like undercover agents wrapped in the innocent and relatively harmless cloak of a non-cancer producing virus, diet, combinations of agents which, on their own, are not carcinogenic, but working together are effective cancer producers, the shuffling of one piece of gene from one part of the DNA to another. All these and many more are what causes a cell to begin to reproduce in an erratic and isolated way, no longer conforming to the needs of the organism of which it is part, but following its own errant path.

The challenge to faith (which, it has to be admitted, is often very hard to face) is to consider whether cancer, although serving no useful purpose in itself, is nevertheless an inevitable part of the sort of world a loving God would have us inhabit when the goal of creation is genuine human freedom. The risk of cancer, flood, plague or famine is the price we have to pay for growing in such a universe. A scientific description of the world has its limits, and within those limits evolution is seen as an aimless process, the laws of physics, chemistry and biology functioning in an environment where chance is a major factor. At a deeper level of understanding, grasped with the intuition of faith, law and chance may be perceived as expressions of the divine will; they lead inevitably, although in their own slow and aimless way, to the emergence of spirit in creatures who can appreciate and respond to a spiritual dimension of the material universe. The laws of science and their interplay with the opportunities offered by chance are the creative method by which the divine ground of the universe brings about the ends it intends.

In evolution, flexibility is the name of the game; openness to change is a major criterion for survival. The cells that make up our bodies and all other living organisms on earth are the active burgeoning product of billions of years of development. If they had not been open to change and modification from within then the elaborate ecosystem of our planet would never have formed, nor would it have thrown up creatures who can respond to a spiritual dimension. Like us, cells are vulnerable, but their vulnerability is a necessary aspect of their flexibility. Only with flexibility can they mutate and transform into cells that build creatures for new niches in the evolving ecosystem, and respond to new opportunities offered by chance. The principle of the survival of the fittest requires that they produce variety and novelty. As heirs to the evolutionary process we have to accept that the cells from which we are made will not always function in our favour. Exploratory, flexible, vulnerable systems are sometimes going to make mistakes. The really remarkable thing is that they work and reproduce so well so much of the time without fault; but it would be as unrealistic to expect that living active cells should never malfunction as it would be to expect a river never to flood its banks.

Science gives a realistic and natural explanation for the behaviour of cancerous cells, for major catastrophes like floods which drown people indiscriminately and for the occasional disaster that strikes earth from the sky when an asteroid collides with the planet. Religious faith should be enhanced by the clarity of this realism. Faith in the twentieth century will understand that occasional floods are inevitable if a watery environment is necessary for life; just as asteroids that fall to earth are a consequence of planet formation, a hazard of the process of world creation. Without the circulation of water evaporating from land and sea falling as rain and rushing down rivers life would probably never have evolved on earth in the way it has. The hydrocycle is part of the dynamics of a living planet. But water can both sustain life or drown it and we have to put up with the fact that too much at the wrong time in the wrong place is a catastrophe. A world in which flood water always flowed away from people or which changed its properties every time it endangered someone's life would be quite unworkable if we are to come to terms with being responsible for our own actions and aware of the consequences. If God always protected us from the unpleasant aspects of nature's laws then life would involve a great many absurdities. There would be no motivation to master the circumstances of our lives; no incentive for moral growth. Wrapped in eternal cotton wool our experiences would add up to no more than an unprogressive and silly dream, an undemanding kindergarten without emancipation or responsibility. Mercifully, the real world is hard and harsh at times, the sort of place in which the human spirit can really grow to maturity.

The freedom God gives us comes with risks such as cancer in the gut or devastating flood, but it also comes with dignity. It is a predictable world in which we can trust the laws of nature to behave with regularity, amenable to interpretation and description. We know where we stand. The rational approach of science is clarifying the limits within which a realistic faith must function. Only in this sort of environment can we pursue our vocation to be human, to discover and make real the image of God latent within us.

Personal tragedies or catastrophes that strike whole communities are no longer believed to be due to the capricious and selfish whims of the gods. Nor do we see them as punishments for sin except in very special cases when people bring disaster down on their own heads. They are not the work of Satan rampaging around the world or of some other deliberately evil force hellbent on ruining God's creation. Catastrophes and cancers, we can now appreciate, are an inevitable part of the sort of reality we inhabit. Death and disaster did not enter the world, ruining its initial perfection, as a consequence of Adam and Eve's sin in the garden of Eden. They were there from the beginning as part of the process of creation. The story of the fall in the garden of Eden and the doctrine of original sin that has been built upon it have a more limited application than orthodox theology has often supposed. Like a parable, the doctrine highlights a truth. The truth is that our various selfishnesses, short sightednesses and evil inclinations interlock throughout families and society as an 'organic system of evil'. But spiritual perfection is not something we have lost, it is a condition we have not yet gained. The garden of Eden is a future possibility, a state of being we all *fall short* of rather than one we have fallen from. The doctrine of original sin contains a true comment on our human situation but through the eyes of science we are bound to read it somewhat differently. It is an analysis of society and its sickness, the prejudices, hates and resentments that afflict it, the evils passed down from generation to generation. We are born, it sometimes seems, into a massive mess and we inherit the blindness and the ignorance of our ancestors. But the fall was not the cause of disease, disaster or death, nor is the pain of childbirth an inherited punishment from Eden (in the nineteenth century women with difficulties giving birth were refused anaesthetics on the ground that they deserved the pain).

Some sort of selfishness is essential for survival and evolution has depended on that for its success. The doctrine of original sin and the Darwinian law of the survival of the fittest can be seen as two overlapping descriptions, though at different levels, of the same phenomenon. We have a self-seeking bias in all of us. The present challenge to faith is not to accept passively the way things are but to respond and grow, to change ourselves and to change our circumstances. A process of spiritual emancipation awaits us by which we may transcend our origins. 'And with the help of my God' wrote the psalmist 'I shall leap over the wall' (Ps 18:29). A new growth into freedom beyond the walls of our heredity, our genes, and our physical environment is offered to us by the Christian gospel. Faith must become action; it too must change and develop. A realistic approach to disease, disaster and death can only strengthen and enhance the growth of the human spirit.

People are not made perfect but are born with the potential for perfection within them. This hopeful emphasis upon future possibilities is not a new theme in Christian theology but has been there from the beginning. Christian teaching in the west has been dominated, it is true, by a 'majority report' which attributes our living in a world of pain and toil to an historic event in the past, the fall: St. Augustine of Hippo (AD 354–430) was one of its leading exponents. A more ancient theology, however, dating back to the second century was developed by one of the church's earliest systematic theologians, Irenaeus (c. AD 130–202), bishop of Lyons.[9] This approach to the human condition offers an alternative perspective; people are to be considered as still in the process of creation. Men and women are made in the image of God as stated in Genesis, but, according to Irenaeus they are not yet *like* God. The blueprint is there within each of us but it is waiting to be realised. This distinction between the image of God and the likeness of God gives a very different orientation to the Christian faith, encouraging the Christian virtue of hope and awakening us to what we may become. Such a view has a high regard for human freedom and responsibility, suggesting that people are the recipients of an invitation to cooperate willingly and freely in God's act of creating the world. Spiritual growth is our calling; an activity only possible when we choose to let it happen (unless a man be born again – from above – he cannot see the kingdom of God, Jn 3:3). This predictable world, explicable in terms of scientific law, is the sort of milieu persons need if they are to grow in freedom; the risks, the pain, the disease, disaster and death, are an inevitable part of it. We can come to grips with it because it is rational; the suffering we sometimes have to endure is the price we have to pay for freedom. Our lives are lived in what has been called 'an environment for soul making'.[10]

We could of course sink back into a sleepwalking mode of life, puppets manipulated by the strings of heredity and environment, but that is not our calling. To be human implies that we become creative moral agents making our own decisions; that we interiorise values so that we can judge for ourselves what is good or bad, beautiful or ugly, life-giving or destructive. We become the authors of our own thoughts and actions, transcending the genes and circumstantial pressures that have dominated our lives until now, and in so doing become more human and at the same time better mirrors of the image of God.

Faith confronted by disease, disaster and death takes a plunge beyond science. Our world is a suitable, if painful, environment, in which we may grow as independent, rational and realistic free beings with love and compassion. It is not enough to analyse what has traditionally been called the 'problem of evil' and to justify intellectually the cruel aspects of creation at the court of a loving all-powerful creator;

suffering has to be lived through, coped with, controlled or conquered.

Science reveals decay and calamity as part of the natural order of things; Christianity's response should not be, as some other religions have been tempted, to sit back passively and accept suffering as an unchangeable fate. A very clear element in our vocation to be human is to seize the reins of circumstance and work for the alleviation of suffering wherever it is found. It must be God's will, and it should certainly be our commitment, that we do everything we can to find ways of identifying, limiting, controlling, and curing cancer. The calling to express our humanity and liberate the image of God within us will need the insights of science and the skills of technology to predict and control the flood and maybe some day to deflect the exterminating asteroid. It will find ways to help our neighbours on this small planet through the great charities, such as Christian Aid, Oxfam, War on Want, Save the Children; it will work publicly through politics or privately through individual acts of love and kindness. The model for this loving concern for people in their distress is the figure of Jesus of Nazareth whose ministry was spent living a life for others, touching, healing and bringing release from affliction.

God does not leave us alone with this painful task according to Christian teaching; he shares and bears it with us carrying a cross in Christ. The whole gigantic evolutionary process that brought us here, from the creation of atoms to the growth of planets and the slow development of life, has all depended upon the regularity of scientific law and its relationship to the opportunities offered by chance. Disease, death and the occasional disaster are not sent by God as punishments but are permitted because inevitable in the method of creation. If God was in Christ then the symbolism of the passion and crucifixion implies that the creator does not sit enthroned outside his world, regretfully contemplating it from a distance, but has somehow allowed himself to become a victim of the painful process.

Christianity can never be accused of an otherworldliness

that tries to sidestep suffering. The word became flesh and nails were driven through it; religious art, it is true, has tended to beautify the theme, turning the cross into a finely worked silver pendant for the confirmation candidate or an ornate piece of polished brass on the parish church altar. In reality the cross was an ugly gallows and when a priest blesses a congregation, he describes in the air, with his hand, an instrument of torture. The incarnation of God on earth might have been as a happy man living a full, charmed life. But it is the particular genius of the Christian conception of incarnation that the story includes shame, despair, rejection and death. God's presence in this world is not like that of a foreign dignitary visiting the squalor of a third world slum, inoculated and vaccinated against infection and disease, well fed and with an airline ticket in his pocket ready to speed away into the sky after saying comforting things and giving a few hand-outs. The God who 'visits' this world stays to the death. Those who know God will know too that 'Christ is risen' and with Christ are those whom he loves.

The extraordinary unity between God and us revealed in Jesus Christ suggests that God knows directly what it is to be the rejected one, the persecuted and the killed. This identification of God with us is remembered recreatively every time bread is broken in the Christian communion meal. At the Last Supper on the night before he died Jesus took bread and broke it saying 'Take and eat; this is my body, . . . Do this in remembrance of me' (Mt 26:26; Lk 19:22, NIV). The symbolism speaks of God's relationship with his world; always self-giving, always wanting to lead the way from brokenness to wholeness from death to life. The world itself becomes the altar upon which the sacrifice is offered, the blood, sweat and tears, the daily toil and pain. Each worshipper should see that the mystery of one's own self is there on the table with the broken bread. The resurrection is the life and the hope one holds to.

Can we dare to hope, then, that this means that God takes the cry 'My God, my God, why hast thou forsaken me?' (Mk 15:34, NIV) unto God again and again, whether it is in the lonely bedroom of the rejected child, on the cancer ward, or in the gas chambers of Auschwitz? I hope so. I believe so.

7. MIRACLES

A miracle has become a thing of the gullible past, in the opinion of many people; inadmissible in an era dominated by a scientific world view. The rationalist tradition, expounded so clearly by the philosopher David Hume in the eighteenth century, would certainly lead us to think so. Miraculous events are banished by this way of thinking to the horizons – if not beyond – of what is probable. Rather than accept the truth of a miracle story, it is more rational to conclude, it is argued, that the witnesses were misled, deluded or plainly mistaken than that the laws of nature, which we come at all other times to respect for their consistency and regularity, have been broken or violated. Hume's sceptical attitude was reinforced by the observation that the average person is thoroughly uncritical about a good tale and will happily embroider it in the retelling. The rationalist's conclusion is that miracles are very questionable events.

The rationalist's rejection of miracles, however, is based upon some very dubious assumptions about the nature of miracles. The word 'miracle' (from the Latin *miraculum*) simply means a 'wonder' and there are many events which will bear the word even though a scientific explanation may be available; the conception of a child by a barren couple, the reversal of a cancer or when a driver walks free from a crushed vehicle. Many of the acts of God in the Bible fall into this category; wonders natural in themselves but seen as signs of God's providence and power. The authors of scripture knew nothing of scientific law and so had no conception of the violation of scientific law. God made the world and caused every detail within it, sunshine and rain, plague, famine harvest and health. St. Augustine in the fourth century, long before the scientific era, said with great farsightedness that 'miracles are contrary to what we know of nature',[1] implying perhaps that with greater understanding we would see them as natural events in the God-given order of things.

The rational sceptic has to admit that the conclusions of science can never be more than probable. Immanuel Kant observed that, 'Hypotheses always remain hypotheses, that is, suppositions to the complete certainty of which we can never attain.'[2] Scientific laws are not eternal facts of nature but are merely the rational generalisations based on experience which scientists make when observing and investigating the world; approximations to truth which may need to be modified or changed in the light of future knowledge. It is important to remember, when discussing miracles, that the 'laws of nature' are not necessarily the hard and fast things we sometimes mistakenly imagine them to be; based on what we know of the world they enshrine our discoveries of regularity and consistency and from them we can make predictions about what is probable but not, let it be noted, about what is certain. The rational sceptic's rejection of miracles amounts to no more than the logical observation that we ought to be cautious about accepting an improbable account of an event – that it flouts the laws of nature – when a probable one is at hand. But even the sceptic has to acknowledge that if there is a God then occasional interruptions of the regular patterns of nature may sometimes happen. Science will not be disrupted by an odd recalcitrant fact but simply has to suspend judgement and await further evidence. If there are therefore such things as miraculous events which genuinely violate the laws of nature, then they will lie beyond the scope of science which looks for order and regularity and which must ignore the possible interventions of the supernatural. The inexplicable *may* happen; as the archangel Gabriel observed to Mary with irrefutable logic, 'For nothing is impossible to God' (Lk 1:37).

Science has its own logical limits and can never rule out miracles; it leaves the question open, arguing that such events are very unlikely from a scientific point of view. A theist, however, will want to go further than this negative defence of the possibility of the miraculous and explore whether there may be positive grounds for expecting the unexpected.

A great divide separates theologians of the New Testament into two main camps, the liberals and the conservatives. A century of textual and source criticism has led the former group to the conclusion that the reader today should exercise a great deal of critical judgement when attempting to digest the message of scripture. She should read between the lines, ask herself questions about why an author records a particular story, be prepared to interpret some events as symbolic or apocryphal rather than as accurate eyewitness accounts of things Jesus said or did. Some miracles will be accepted by adopting this approach to the gospel stories; others will be rejected. Conservative theologians on the other hand will tend to accept the evangelists' account of Jesus' life and ministry as essentially historical in content, recording the reliable and trustworthy memories of the disciples; if they say Jesus walked on the water then without question that is what he did.

A useful distinction can be made between the nature miracles of Jesus and his healing miracles. The former include such stories as the stilling of the storm (Mk 4:35–41), the walking on the sea (Mk 6:45–52), the transforming of water into wine at the wedding feast at Cana (Jn 2:1–11) or the feeding of the five thousand (Mk 6:30–44). All of these involve clear violations or interruptions of the laws of nature as we know them; they manipulate the winds, overcome gravity, interfere with chemistry or multiply matter. The possibility of freak inexplicable events such as these is not entirely ruled out by a rational approach to nature; the occasional miracle does not come under the purview of science. A conservative reader of scripture can feel secure in the knowledge that it is perfectly logical to accept the historical accuracy of the nature miracles and to believe that they were part of a disclosure of divine power and signs of the kingdom of God on earth. They are an aspect of revelation and tell, through Jesus, something of the character of God.

A liberal reader will also interpret these miracle stories as signs but will not feel the need to insist that they actually happened. They are not news items about public events but should be read as symbolic stories and not history. As symbols they tell us of the profound belief of the disciples that God was in Christ. The stilling of the storm is a powerful image of God's control of the forces of nature; the God who is revealed through Jesus is the same God who created the heavens and the earth in Genesis and whom the psalmist hailed as Lord of nature – 'Thy way is in the sea, and thy paths in the great waters' (Ps 77:19). The story of Jesus providing wine unexpectedly at a local wedding is a parable proclaiming him Lord of the messianic banquet. Rabbinical literature of the time used the image of a great feast to portray the messianic age which many Jews believed was about to break in on the world. The story must have been reinforced by similar legends of the Greek god Dionysus who performed the same water-into-wine magic. The important point to bear in mind is that this sort of liberal interpretation of the nature miracles is saying nothing less than its conservative cousin about the unique revelation of God in Christ, a man. They are symbolic stories which tell of a deep truth.

Healing miracles are a very different order of event from nature miracles; an age which has grown accustomed to the concept of psychosomatic illness ought to have no difficulty in accepting the general accuracy of the accounts of cures that Jesus produced in those around him. Like other healers of his day – and since – he had powers, or was able to release powers in those who came to him, which made the handicapped whole. Here at least there can be substantial agreement between liberal and conservative theologians for we are no longer dealing with the sort of events which a sceptic might think fly in the face of reason. It is a commonplace observation that people may make themselves sick with anxiety; that stress may be one factor in the cause of cancer; that the over-worked entrepreneur may get stomach ulcers. The obverse is equally true, that physical healing may often depend on spiritual wellbeing.

The stories of Jesus' healing miracles while historical at

core also have a symbolic depth to them. The author of Mark's gospel, for instance, is careful to place the account of the gift of sight to the blind beggar Bartimaeus at the point in the narrative where Jesus is opening the eyes of his disciples to both his and their destinies (Mk 10:46–50). And his answer to John the Baptist's question 'Are you the one who is to come . . .?' (Lk 7:19, JB) suggests that he understood his ministry to be more than just a matter of good works: 'Go back and tell John what you have seen and heard: the blind see again, the lame walk, lepers are cleansed, and the deaf hear, the dead are raised to life, the Good News is proclaimed to the poor . . .' (Lk 7:22, JB). He was saying, in other words, that the signs of God's kingdom, hinted at in the book of the prophet Isaiah, were being realised in Galilee. Time and again the point is made, in the gospel accounts of healing miracles, that Jesus could only set people free from their illnesses when they had faith or when he had released them from their sins by forgiveness. In the universe we have been describing in these chapters, the human spirit is an emergent quality of chemistry; body and soul are two aspects of one organic phenomenon. Healing miracles violate no principles of science for they are in accord with the psychosomatic laws of nature. There can be no rational objection, therefore, to the belief that Jesus was able to heal some lepers by breaking social taboos and touching them, that he gave strength to the bed-ridden, set loose the tongues of the dumb and opened eyes. It would be thoroughly unscientific to deny that healings can sometimes happen in this way.

In the human spirit there is an untapped force for good; faith, love and forgiveness are themselves creative powers which can work miracles. To this extent miracles are natural in that they are rooted in the mystery of the human individual. Even some of the so called nature miracles of the Old Testament may be interpreted in this way. The exodus, passover, Red Sea crossing and wandering in the wilderness is the great archetypal miracle of scripture (Exod 12–17). God interfered in history, manipulating the pharaoh and the weather so that the Israelites might make their exit and be led to their own promised land, a land flowing with milk and honey. There was nothing unnatural about the miraculous way they survived in the desert: quails still flop exhausted from the sky in spring after their migratory flight across the Red Sea en route to Russia; soldiers during the second world war rediscovered that water can be obtained by breaking through the crust of rock on the sides of a wadi; 'manna' – a sweet wafer like substance, the glucose full secretion of an insect inhabiting tamarisk trees – is still gathered by bedouin Arabs of the Sinai; and even the symbolic Red Sea crossing (God rules the chaos of the waters) is said in the Exodus account to have been due to the blowing of a strong wind. The *real* miracle in the Exodus story is what God worked through a person, the once-exiled Moses. Fearful for his life (he had killed an Egyptian) and fearful that his own people would not listen to him he nevertheless returned to Egypt fired with the belief that God could revive a dead nation or lead the Israelites to freedom. He had seen the bush burning in the desert and knew it would live sprouting again with the next rain; if God could do that naturally for a bush then God could surely do the same for the chosen people. The Israelites saw the hand of God in the story of Moses and in all the attendant circumstances which worked for the Hebrews as they became refugees. It was a wonder, the greatest miracle known to them, and it became central to their account of history and their relationship with God. The exodus strengthened their faith, demonstrated God's presence with them and opened up new and creative possibilities for their future. Miracles, we may learn from this piece of history, work through people.

The Exodus miracle, the resurrection of a dead nation is paralleled in the New Testament by the account of Jesus' resurrection on Easter Day. The empty tomb is the most significant of all the miracles recounted by the evangelists. It is the climax of the story in each of the four gospels and is the heart of the Christian message as summarised by Paul in his first letter to the Corinthians (written many years before the evangelists wrote the gospels). The claim that he truly died on the cross and really came alive again is impossible to verify as an historical fact although there is no doubt at all that it was this which essentially gave the title Good News to the earliest proclamation of Christian teaching. The first followers of the way believed it passionately to the point of death; many willingly became martyrs for the faith.

An interpretation of the Easter story is not without its difficulties; once more we find twentieth-century theologians divided. 'Flesh and blood cannot inherit the kingdom of God' wrote St. Paul (1 Cor 15:50, JB) and so how does a thirty year old wounded Jewish body fit into what we believe about heaven? Our resurrected bodies, we are led to believe, will be as unlike our mortal bodies as plants are unlike the seeds from which they grow (1 Cor 15:35–38). If Jesus really walked out of the tomb alive, his dead mutilated body revivified, then what did he do with it when the resurrection appearances came to an end? In the creed we state that after his resurrection Jesus ascended into heaven and sat down at the right hand of God; both statements use symbolic imagery. If we are happy with a theology which interprets these articles symbolically and not literally then why not treat the resurrection story in the same way? It is perplexing.

Many Christians who believe deeply in the truth of the resurrection nevertheless find themselves asking these questions. The problem is not made any easier by the gospels themselves which contain conflicting and somewhat inconsistent details. Instead of a clear account of what happens we have a number of brief fragments many of them defensive in nature; he was not a ghost – he ate fish; it was not a plot by the disciples – they also disbelieved the resurrection at first; there was no mistake about who he was – they saw the marks of his crucifixion; he really did die – the Roman soldiers checked and then put a guard on his tomb. The oldest manuscripts of Mark, considered by many scholars to be the earliest gospel, recount no resurrection appearances at all but end with the empty tomb and a gasp of wonder. Luke gives the impression that Jesus appeared to his disciples a few times in and around Jerusalem before ascending into heaven after forty days, while John paints a contrary picture in which Jesus appeared

172   UNIVERSE

to them again after they had returned to their trade of fishing
on the sea of Galilee. Paul then confuses matters even more
by implying that his own vision of Jesus on the road to
Damascus, years later, should be included in the list of
post-resurrection stories (1 Cor 15:8).

Are the resurrection stories 'faith legends' expressing a
spiritual truth, as some liberal theologians imply or did Jesus
really physically and historically rise from the dead? The
question is not by any means a new one. Gnostic writings that
come to us from as early as the first century AD, written by
contemporaries of the gospel evangelists, treated the resurrection as a spiritual rather than a physical phenomenon.[3]
Their minority opinion was judged to be heretical by the
orthodox majority and so their view never reached the pages
of the New Testament.

A twentieth-century Christian who feels that his or her faith
must necessarily be grounded in reason need feel no scepti-
cism about the powerful truth contained in the Easter liturgy
'Christ is risen — He is risen indeed'. Either he or she will
adopt the liberal theologians' interpretation of events and be
content to believe that the story conveys a spiritual truth about
death and eternal life, or will without any hesitation accept
the historical accuracy of the resurrection stories with their
various attendant problems. Frequent interruptions of the
laws of nature by God would make the world unworkable as a
place for the growth of independent autonomous souls. We
need the regularities of a predictable world if we are to take
responsibility and exercise moral choice. But the open nature
of scientific description means that the possibility of an occasional miracle is not ruled out and one extraordinary inter-
ruption may act like a tiny piece of yeast which leavens the
whole loaf. By raising Jesus from the dead physically and
demonstrably God revealed a truth about our mortality and
what may become of it. Sin is forgiven, death conquered
and resurrection is our destiny – but only if we accept the
invitation, for even in this matter God does not overrule
the freedom God gives us.

The same criteria may be used for evaluating that other

great miracle of Jesus' life, the virgin birth. The nativity stories do not have anything like the same significance as the resurrection does for early Christian preaching. Neither St. Paul nor the author of Mark make any mention of the virgin birth, they may not have known the story and at any rate seem to have believed that the gospel they taught was complete without the doctrine. Moreover the author of the fourth gospel felt able to expound his theology of the incarnation without any reference to how Jesus was conceived. The birth stories in Luke and Matthew are our sole source and they show distinct signs of moulding by the purpose of the writers and the expectations aroused by their reading of the various proof texts from the Old Testament believed to foretell the manner of Christ's coming into the world. The Greek translation of the Hebrew Bible, in use at the time, suggested that the mysterious Immanuel figure whose birth was predicted by Isaiah (Is 7:14) would be born of a virgin; the original Hebrew implied no more than that she would be a young girl. The miracle of his birth might then be seen, by a liberal theologian, against the background of other biblical God-sent nativities. Did not the three mysterious young men who met Abraham beneath the oak tree at Mamre say to the aged and doubting Sarah 'Is anything too wonderful for Yahweh?' (Gen 18:14, JB).

The tentative nature of the evidence however does not force us to abandon it. The doctrine of the virgin birth may have been a truth which the church discovered at a later date and it is reasonable to suppose that it is the sort of thing Mary would only have revealed late in life. However, given that the earliest formulations of the gospel managed well without the doctrine, it is a pity that in the late twentieth century it has become something of a shibboleth of Christian teaching. Petitions are circulated to denounce and demand the resignation of bishops who view the doctrine as symbolical and not biological.

It is argued by some people that the theology of incarnation demands a virgin birth; but the question can never be settled definitely. It is a foundation dogma of Christianity that the word became flesh, but a Christian may find that he or she has no choice but to remain open minded on the matter of 'How?' It is argued that when a child is conceived a new person is created and since Jesus was not a new person but the word made flesh then normal conception cannot have been the mode of his entry into this world. However, his whole physical being, from the skin of his head which one day was to be pierced with a crown of thorns to the soles of his feet which would be mutilated by nails, was grown from the DNA inherited from his mother Mary; he looked Jewish because she was Jewish. Jesus' biochemistry was the same as that of the rest of the human race in all respects. If the DNA in Mary's ova could be activated to produce the Christ child then why not in combination with the DNA of Joseph too? The unity of divine and human proclaimed in incarnation and the manifestation of spirit as a dimension of matter is already a miracle the 'How?' of which may ever remain beyond the analytical powers of our minds. Whatever the manner of his birth Jesus' life and death spoke eloquently of God, and Mary may still bear the title *Theotokos*, 'Mother of God'. By insisting on the biological details of the virgin birth we may be missing the point of a much profounder miraculous event. Jesus was not a superman cuckoo, a spiritual alien planted in a girl's womb; he was a revelation of God through the mysteries of human biochemistry. Rooted in the gene pool of humankind, which has been raised from the dust by evolution, God offers us the way to eternal life.

The pressing question about miracles is a contemporary one; can Christians who have been educated to think of the world in scientific terms still confidently expect the unexpected wonder in their lives? Are intercessory prayers worth the time and energy? or is everything predestined by scientific law, a sort of infallible cosmic train timetable of events with which God will never interfere? There certainly has been a shift in our thinking about the mode of God's relationship with creation. In the 1662 *Book of Common Prayer* it was thought perfectly reasonable to pray 'send us we beseech thee such moderate rain and showers . . .' as though by stirring a divine finger in the atmosphere, God could create the odd cyclone, disperse the blocking ridge of high pressure, or fan with a breath of rain-bearing monsoon up from the ocean to drought afflicted lands. No doubt God could — for God nothing is impossible — but we concluded when considering catastrophes that it is not God's way to interfere with the laws of nature. Christians today are much more likely to think in terms of praying for guidance and strength, for charity and mutual support in the face of disaster rather than for a change in the weather. We have become more realistic and somewhat stoic about nature; we favour practical action, reckoning that God wants us to take some responsibility and behave with intelligence. Londoners built a costly Thames barrier at Woolwich to prevent the city from being flooded by the North Sea, rather than place their security in waves of prayer organised across the home counties every time the tide runs high.

Miracles work through people. God's will for people, as revealed through Jesus Christ and proclaimed in the gospel, is to set captives free; to liberate the image of God in men and women. This is why miracles can realistically be expected to happen. There are more possibilities for novelty, for action or inaction, facing any given moment than we can ever imagine. The world of people is not as predetermined as a robot factory but offers many opportunities for a creative response; there are many ways that the present moment may work itself out into the future. Just as at the level of subatomic particles there is a principle of 'indeterminacy' first described by the scientist Heisenberg so by analogy there is at the human level an 'indeterminacy' about our futures, both physically and spiritually. In a way, all novelty and creativity is miraculous and it is only our dulled sense of familiarity with the world that blinds our eyes to this vision.

Science does not dispel the wonders of this world or hinder our appreciation of the miraculous; nor does it banish God to the peripheries of the universe. God is the eternal presence in the process, the creative ground of the universe; the laws of nature and opportunities offered by chance express God's will. Human beings are the conscious heirs of the evolutionary process and are able to utilise its laws and opportunities in a way not open to any other creature known to us. The reality into which we are born, wake up and grow, offers us more, however, than just the privilege of free choice (a privilege we may, out of habit, laziness or ignorance, never realise). It presents us with the invitation to share in God's creative powers through faith. The creative power of love is as much a law of nature in the realms of the human spirit as electricity is part of the dynamics of a thunderstorm. In a universe rooted in God, miracles amongst people are natural.

'Chance favours the prepared mind' wrote Pascal; and in a slightly different vein William Temple, Archbishop of Canterbury, said of the minor miracles of life that 'when I pray, coincidences happen. When I cease to pray they don't'. It is part of our humanity that we can be physically transformed by love, and that our personalities are expressed most fully through the use of our freedom of choice. God, working with and through Moses, turned a national tragedy into a new beginning when he worked a miracle and liberated the Israelites from slavery; Jesus, the incarnate word of God, was able to free people from their individual sicknesses through the power of faith, love and forgiveness; the Holy Spirit, God's creative power experienced within as a personal force, broke down the barriers of fear and suspicion at Pentecost and opened up channels of communication where before there had been a babble of misunderstanding. The liberating effect of the Spirit on the lives of the disciples after that first Whitsunday, described by Luke in the Acts of the Apostles, spread through society like ripples across a pond; the healings performed by Jesus were repeated by his followers.

With faith and prayer miracles can work through people today. Individuals can experience physical healing and communities can be brought together. We must believe that where there is division, fear and suspicion there can be respect and trust whether in Belfast or Hebron, Johannesburg or Bombay.

There is nothing in a scientific description of the world

which robs faith of its duty to expect the unexpected with hope. A rational and intelligible order still leaves room for miracles. Indeed science may help liberate miracles from the land of mumbo-jumbo from the conjuring tricks of some fakirs and shamans and put them centrally in the realm of faith and the creative powers it makes free. God does provide for his children but not by overruling our freedom or waving a magic wand from the sky. God provides an environment in which we may become compassionate co-workers with him and by summoning us to greater acts of humanity. Faith liberates the creative powers of love and miracles have a rational place in a human universe.

8. FAITH AND FALSIFICATION

Religious statements about reality are not in the same category as the facts of science. The articles of faith can never be tested with the sort of detached rigour to which we subject scientific theories. Nevertheless we want to know whether or not they are true.

Religious beliefs have become the ghostly ephemeral cousins of scientific 'facts'. If they are to retain any value then they must discover a new firm place in our world. The scientific method has cast the thinking of western culture into a misleading dualism of subject and object. It has of course been a necessary and fruitful part of scientific procedure to work with the concept of an objective reality 'out there' which can be investigated, analysed, checked and double checked; it has paid immense dividends. Our knowledge and potential control of the world has expanded as a result. But in the process the importance and significance of the observer came to be neglected. Religious beliefs became private value judgements and were reduced to being 'merely' subjective, of personal interest only, carrying no weight in accounts of the real world.

The balance has been redressed in quantum mechanics at least. In the realm of subatomic particles the observer, (that is the scientist doing some piece of research) is an integral non-isolatable part of any experiment. The act of observation is part of the phenomenon observed; the very act of looking, whatever instruments are being used, affects and influences the thing being looked at. There is no objective world 'out there' which the scientist can describe without reference to himself or herself. This insight of quantum mechanics is even more true when it comes to the contents of faith. The truth or falsity of religious beliefs can never be considered in separation from the people who hold them. It is only in the relationships between me and my personal history, my environment and society, and the basic mysteries of my existence that my beliefs can be either true or false. Faith will take a step beyond quantum mechanics' discovery that the observer has a significant role and will develop the view that what is revealed to the whole person, his subjective view of reality, can involve insights as profound as any description of the world made by science. A living, loving, vulnerable human being is an instrument evolved by nature for perceiving truth.

The human brain is the most complex phenomenon amongst the stars, and the spirit that emerges with and through it is the most remarkable quality of matter in the known universe. It is far more refined and sophisticated than any scientific instrument devised by technology. It is to be expected that such a wonderful entity will have important things to say about the world it inhabits and observes. Scientific truth is only one aspect of the universe; there are other truths with which religion deals. A human being's grasp and appreciation of reality through myth and symbol makes a real contribution to a coherent vision of why we are here at all. The subjective opinion about religious claims which begins with 'it seems to me . . .' is undervalued and neglected at our peril. It is natural for us to ask questions about why we are here and to seek meaning; there is no reason to assume we will be disappointed when it comes to finding answers or that our minds will merely generate delusions. We were evolved to respond to and recognise the truth.

Which subjective opinion? That is the question. An atheist's view of the universe may be every bit as coherent as that of a religious believer; between religions there are immense contradictions. Who is right, and can religious statements be checked, tested or falsified in the way demanded by a scientific theory? Karl Popper, author of the falsification principle, exposed the weak position of those who cling tenaciously to their pet theories avoiding falsification whatever the evidence: 'The wrong view of science is revealed in the craving to be right'.[1] Anyone proposing a theory about the way things are should be prepared for attempts to be made to disprove it. What would count as evidence against it showing the theory to be false?

Astrology is a good example of a theory or cluster of theories which seem to avoid falsification at all costs. The claim that the fortune of each of us lies in the stars purports to be empirical but is really pseudo-science. A basic proposition is that your day to day experiences and character are influenced by your birth sign; that the stars rule your life. Many well educated people are still unclear about the distinction between the science of astronomy and the art of astrology, and more columns in journals and newspapers are devoted to 'what your stars foretell' today, than at any time in the past. Some research, indeed, has gone into testing the link between birthdates and professions with interesting results. It appears, for instance, that university teachers are more likely to be born in May than in any other month and that people in top jobs tend to have birthdays in the spring than in the autumn.[2] A close scrutiny of sun-sign astrology however does little to back the claims that our destinies can be read in the stars. Seasonal or climatic factors may explain the trends just mentioned more rationally and satisfactorily.

The usual vague way of testing a daily horoscope is blithely to accept those predictions which seem to be supported by experience and to forget about all the rest. It is remarkably easy to read a well worded description of a Virgo character for example and to assent to all those elements which seem to suit the character in question. A theory supported by such a selective use of evidence deserves to be called superstition rather than science. But the biggest piece of evasion of falsification in astrology is quite startling.

The foundations of birthsign astrology have shifted to such an extent that the whole edifice should have come tumbling down long ago. When the rules of astrology were laid down over two thousand years ago a fixed point of reference was the spring equinox. The spring (or vernal) equinox is that time in the year when the sun in its apparent path across the sky crosses the celestial equator from south to north. The location in the heavens where this happens is known as the first point of Aries because it used to lie in the constellation of Aries the Ram. Modern day astrologers still take this to be so despite the fact that due to a phenomenon called the precession of the equinoxes the first point of Aries has moved backwards into the constellation of Pisces the Fish. Every sun sign is nowadays a whole constellation out of step.

A brief explanation will help to clarify this oddity. As the earth orbits the sun, the sun appears to follow a regular path through the heavens during the year. This path is called the ecliptic and passes through the twelve constellations of the zodiac. A person is said to be a Virgo-type if the sun is in front of that constellation on the day he or she was born. Now the earth's axis is not perpendicular to the plane of the earth's orbit; if it were so the sun would remain above the equator all year round and all seasons would be the same. As it is, we have seasons of summer and winter because for half the year the northern hemisphere, say, is tilted towards the sun which is then high in the sky, while for the other half of the year it is tilted away from the sun. Spring and autumn lie between the two extremes. The first point of Aries is the spot in the heavens where the sun is in spring as it edges higher up the sky giving us longer days and shorter nights. So far so good; but now for the problem. The earth's axis is not steady but is subject to a slow wobble, just as a nursery top while spinning will also begin to sway drunkenly as it moves around the floor. The earth's axis takes 26,000 years to complete one drunken sweep. The result? – every couple of thousand years the point where the ecliptic crosses the celestial equator slips back a constellation. Everyone who thinks he or she is an Aries is a Pisces (and if surviving into the next century will become an Aquarian); Virgos are really Leos; Leos are really Cancers and Cancers are Geminis and so on. Astrology is stuck with an old timetable and blissfully ignores this major shift of the sky.

Despite all the evidence which might be taken to falsify astrology the belief that the stars have a real influence in our lives remains for many people a captivating delusion. This need not matter. We may conclude on reflection that astrology is a form of light entertainment not taken too seriously by the majority of people; or even that it is a form of harmless therapy. It will only do damage if it undermines a person's sense of freedom and responsibility or if it raises false hopes. The significance of the current popularity of sun-signs is that it expresses a deep seated yearning in twentieth-century people to find order in the cosmos and some sort of meaning in life. Perhaps its appeal lies in its having an affinity with the truth; that we *are* a product of the stars and that our destinies cannot be isolated from the material universe which has given birth to us.

Religious statements are harder to test than any of the hypotheses of science or the fanciful claims of pseudoscience. They purport to be insights into reality but their truth lies at a different level from that of scientific description. They have a close kinship to value judgements, attitudes or opinions and a different sort of criteria from scientific methodology are needed to examine their truth or falsity. One would not, for example, examine the claim that Bach's Double Violin Concerto is very beautiful by measuring the length of certain strings or by recording the frequencies of particular notes.

The immense variety of religious claims, many of which seem to contradict others at a fundamental level, constitutes for some people a disproof of their truth. The British philosopher Antony Flew has commented on their apparently basic unreliability; 'religious experiences are enormously varied, ostensibly authenticating innumerable beliefs many of which are in contradiction with one another'. The orthodox Jew may expect nothing after death, the Hindu anticipates another life reincarnated on earth, while the Christian hopes for heaven; at the heart of everything there is a void, sunyata, according to the Buddhist, but for the Christian it is God. The perplexed atheist may be forgiven for concluding that religious beliefs have no more truth in them than do the patterns our minds may idly see in the clouds on a summer's day; delusions all of them.

Japanese families in the sixth century AD, fascinated by the art, religion, philosophy and culture of China decided to test the Buddhist faith of the mainland. They put their Shinto gods to one side and worshipped images of the Buddha for a season; plagues abated, the weather improved and Buddhism was established as a worthwhile thing. The new religion began its long history of coexistence with the indigenous faith. This sort of 'pragmatic superstition', as one scholar has called it, was much more possible in the east, where attitudes to religion are more eclectic, than it would have been in the west. (Though a similar contest, enshrined in our tradition, was held by the prophet Elijah and the prophets of Baal on Mount Carmel.) The idea that religious beliefs might be tested is not new, but one would expect an enlightened Christian of the twentieth century to approach the business with more spiritual sophistication.

A paradigm shift has affected the thinking of many Christians as they examine the truth claims of religious beliefs. It involves a radical switch of thinking of the sort identified by Thomas Kuhn as characterising scientific revolutions.[3] Such a Christian no longer assumes that the body of Christian doctrine constitutes an exclusive statement of what is true. This does not mean that he or she doubts the Christian way any more than did our ancestors; it is simply that the Christian doctrine is seen within a different framework.

We live on one world in one universe based in and emerging from one divine Ground. 'We believe in one God the Father Almighty, Maker of heaven and earth, and of all things visible and invisible', begins the Nicene creed. The Logos, the word of God of which creation is an expression, is not only revealed through the pages of the Bible; inspiration is not limited to the prophets of the Old Testament or the writers of the New Testament. A Christian who has undergone this paradigm shift will expect to find deep insights in religions other than one's own and will reckon that the opposite of one profound truth may sometimes be another profound truth. A dialogue between different traditions will develop, one in which the participants will seek to learn from each other in humility, sharing inspirations and together acknowledging the Mystery.

The Christian way can no more be tested by merely sitting and thinking than a scientific theory can dispense with experiment and observation; both need the light of experience. 'The Christian ideal has not been tried and found wanting', wrote G. K. Chesterton, 'it has been found difficult and left untried.' The seeker has inevitably to get completely involved in the testing and there can be neither any short cuts nor neat definitive experiments which will establish the objective truth of Christian beliefs. You can't learn to swim without climbing right into the river, bracing the cold, letting go the bank, immersing yourself and getting a grip on your fears.

There is a two-way movement in the Christian experience of religious truth, one directed inward and one outward; each is sustained and strengthened by the other. The first movement is inspirational, the activity of the Spirit blowing like a fresh breeze into our minds and hearts; the second is our willing response. The universe is rooted in meaning because it is grounded in God. There has to be an attentive, willing and relaxed assent on our part for that meaning to break in upon us. The language of this revelation which comes to us is in the language of symbols. Religious symbols have their own unique authority and power for conveying truth; symbols such as incarnation, light, unity, the cross, resurrection and so forth. In them and through them the creating Spirit, always present and waiting for our receptive wakefulness, makes a movement into our lives. The image of the dispirited frightened disciples locked in an upper room weeks after Jesus' crucifixion is the archetype of this divine movement in our direction. Like a rush of mighty wind the Spirit blew into their stuffy self-inflicted captivity and settled like a flame of new fire on each of them (Acts 2:1–4). Exhilarated, refreshed and remotivated they threw open the doors stepped out into the world; and that is the second movement intrinsic to the Christian way, the personal act of will, the deliberate journey each of us begins to make.

The appreciation and contemplation of symbols requires

that time be made available for meditation and contemplation. Only then can they be known by the fruits they bear in our lives. We are born into and grow up in a reality which is rooted in mystery, a world in which spiritual insights are every bit as real as scientific observations. We are products of a spirit-matter process and have evolved from nature as creatures uniquely suited to the comprehension of meaning. Flycatchers evolved to fill a niche and catch flies, bulky whales developed massive streamlined bodies for the ocean deeps, *Homo sapiens* has been evolved among other things to be alert to symbols, he is an organism which can be tuned to the truth. It is because we have emerged from a spirit-matter process, and not from a meaningless universe, that some symbols find a resonance in our minds and hearts. They ring true.

God is a symbol: the faith of the west is built upon this rock. When I say that God is a symbol I do not mean it in the sense implied by Don Cupitt, Dean of Emmanuel College, Cambridge.[4] He argues that the Christian doctrine of God is just Christian spirituality in coded form; God is a fiction, giving focus to all our spiritual hopes and ideals. God, in my view, has a more positive reality than that. The symbol suggests to me that at its roots reality is personal not impersonal and that the world is the product of will and intention and not the haphazard absurd result of blind accident (though haphazardness in evolution has, as we have seen, its own significant place in the process). The gospel invites me to respond to God as I would to a personal relationship and not simply to cultivate my own solitary spirituality. The God-symbol as a statement about reality cannot be isolated from my attitude to reality; it is tested, corroborated or falsified by the way I live. Philosophers have concluded again and again that no amount of logical thinking will establish the existence of God; the Muslim philosopher-turned-mystic al-Ghazālī demonstrated in his own life the importance of the distinction between knowing *about* God, and knowing God.[5]

The symbols which arise out of the facts of Jewish-Christian history are central to a Christian vision of the world. The exodus-symbol speaks of liberation; the symbol of the city of Jerusalem evokes the vision of the city of God. In the gospels the incarnation, that miraculous unity between God and the human in Jesus, is fundamental to Christian theology. His death gave the world one of the most powerful symbols of all time: the cross. So much of human experience is caught up in and resonates with the image of the cross; despair, desolation, rejection and death, all are worth living through for what lies beyond them. There is a wealth of riches in this one symbol alone, but it cannot be extracted from the gospel and isolated; it is integral to the story that was enacted in history. The resurrection is another facet of this same gemstone of truth and is an equally fruitful source of grace and meaning. In the Christian view of the world a whole constellation of symbols cluster together providing a coherent vision and source of revelation. But the Christian today will not only find meaning through the orthodox doctrine of the church. Symbols from other sources may be equally evocative.

The Buddhist symbol of the lotus, for example, is an image of enlightenment and speaks eloquently of an experience to which we might aspire. Gautama, who became the Buddha – 'The Enlightened one' – had a thoroughly pragmatic approach to spiritual truth. The right way, which he mapped out for his followers, known as the Noble Eight-Fold Path, has survived repeated testing.

'Come and see' he invited his contemporaries whether farmers or kings, merchants or statesmen. He was a teacher of demonstrable empirical truths, he insisted, and no one should believe what he taught simply because *he* taught it. He indicated a practical method which any man or woman could put to the test. Enlightenment, the goal, was as much a fact of the real world as the blooming of a lotus. The lotus grows in the mud and murk at the bottom of the pond, it pushes its way up through the water towards the light until it reaches the surface. In the freshness of the air, the light and warmth of the sun it bursts open and blossoms. Compare this symbol with the saying John attributes to Jesus 'Except a man be born again, he cannot see the kingdom of God' (Jn 3:3, AV). Both images, rebirth and blossoming, speak of the same need for a new spiritual wakefulness; something we can prepare for by meditation and contemplation but something which in the end happens *to* us.

Poets provide symbolic images that speak of matters of faith too; indeed it has been said that there is a close kinship between poetry and theological language. There is a line for example, in T. S. Eliot's *Four Quartets* which I have returned to many times for the way it evokes what I believe to be true, 'The still point of the turning world'.[6] Everything in the cosmos is moving, falling, turning; the earth spinning on its axis daily, orbiting the sun yearly, drifting eternally amongst the stars of the Milky Way as the falling galaxy itself revolves slowly about its centre. The discoveries of astronomy are matched by the theory of evolution in setting us loose in a world process of change and movement. We live in a turning world, but it is a world rooted and grounded in stillness. God is the unchanging life at the heart of the universe and at the heart of every atom, the quiet field in which all things have their orbits, their motions and their transitory agitated existences. Between the stillness and the restlessness creation happens.

We dwell here, each of us, at the interface between two mysteries, the ordered universe which surrounds us and our own inner world of the spirit. The poetic and theological languages of faith can help us comprehend this strange position in which we find ourselves; they can make coherent sense of the 'without' and the 'within' of things. The turning world of T. S. Eliot has a local reference similar to the cosmic, one of more immediate concern to our faith. The pressures and demands of family life, of business and the daily enterprise of survival, the encroachments of old age, disease and death, can at times become a maelstrom of insecurity, fear and despair. Churning anxieties may invade my inner being but experience tells me that there is a still point to be found too in this turning world and at the still point there is life.

Faith *can* be tested. The symbol of changelessness lying at the heart of change, stillness as the pivot of movement, can become the vehicle of a real experience. An anonymous poet, the psalmist, found it to be true over two millennia before T. S. Eliot, 'Be still, and know that I am God' (Ps 46:10, AV). The biblical commandment to keep the sabbath holy as a day of rest enshrines the need to be still in a law but perhaps ties it too securely to one particular day. It need not be Saturday or the Christian Sunday or even every seven days. The value recognised by the commandment is that people need to stop regularly to experience rest rather than restlessness; to think, to pray, to meditate and to let the healing life that is there in all situations come to the surface. But stopping in order to let stillness become real has to be supported by a prepared mind.

Peter Medawar writes of the scientist that by his research and reading 'he has made himself discovery prone'.[7] In an article about the mechanics of scientific discoveries, he argues that they cannot be predicted or commissioned but neither do they simply come out of the blue accidently: 'A scientist is a man who by his observations and experiments, by the literature he reads and even by the company he keeps is putting himself in the way of winning a prize . . .' This preparedness of mind is not an attitude that can be adopted casually but requires the investment of immense amounts of effort; it is 'worked for and paid for by a great deal of exertion and reflection'. Similarly the person who follows the Christian way by regular reading and meditation on the lines of the Lord's prayer, for instance, and by giving time for stillness, has made herself 'inspiration prone'. Unexpected riches may begin to bubble up. A fractured, fragmented life will begin to come together and she will experience for herself how the gospel can set a person free; liberation theology at first hand!

Many of the most important Christian teachings are not directly statements about God or Jesus at all. Doctrines such as that 'God is one' or that 'God was in Christ' are fundamental to the gospel, but much that is of value in Christianity is in the form of guidance about how to live. It recommends a certain style of life. Here is an obvious area where the practical tenets of faith may be tested. Many mechanisms we have designed involve 'feedback' and adjust themselves using the information they collect as they function. A central heating system with a thermostat works on this principle. Human beings even more than the mechanical systems they build can become self-correcting entities. We can register and monitor the effects upon ourselves and others of adopting a particular attitude. In fact we learn from feedback in this way all the time; we are conditioned from childhood by praise and blame for instance. All I am suggesting is that there is ample opportunity for extending this approach to life by making it more self-reflective and aware. Some sports coaches use a similar method which they call the 'inner game', encouraging the tennis player, say, to become more aware of the feel of a shot when it goes right and noting how it felt when it went wrong.

Three paradoxes of faith may be tested by this 'inner game' feedback method: the paradox of commitment, the paradox of forgiveness and the paradox of suffering. There are others I could have chosen but three will suffice to show that the Christian way can only be tested by living it wholeheartedly.

The paradox of commitment is that liberation comes with self-giving, just as you think you have tied yourself down you find the reverse is true. When a person commits himself or herself wholeheartedly to another as in marriage, or to a community to love and serve it, the experience is not one of being tied up, ensnared or trapped but is of a new sense of freedom. This paradox is rooted in the fundamental belief that we achieve liberation by serving God; or as the second collect in the Office of Morning Prayer has it 'O God, who art the author of peace and lover of concord, in knowledge of whom standeth our eternal life, whose service is perfect freedom . . .' Such a freedom can only be known to those who have made the commitment by a deliberate movement of the will – it may take years to realise – and their experience will be powerful corroborative evidence for this particular claim of faith.

The paradox of forgiveness is stated in the Lord's prayer; you cannot receive forgiveness unless you have given it. It might seem from this that it is merely a dry contractual matter; obey the rules and God will treat you well; one more example of how Jesus made his followers accept ideals which were even more demanding than those of the Pharisees. But the whole spirit of the gospel implies something different. The teaching about forgiveness is not so much a command as a piece of good news. Believe it or not, it suggests, despite the cancerous qualities of hate, jealousy or resentment, it is actually *possible* to forgive the person who hurt you. Your 'enemy' may be someone close to you – husband, wife, brother or mother – and living with them unforgivingly may constitute a great burden. The paradox is that when you *do* forgive someone for hurting you, *you* are the one who will feel released, freed to discover the lightness of the kingdom of heaven (and of course so may they). Only by truly forgiving can we taste the fruits of this teaching and it may take a long time; 'not seven times but seventy times seven' as Jesus answered Peter (Mt 18:22).

The paradox of suffering is revealed through the mystery of the cross. It may often seem that 'man's life is a cheat and a disappointment';[8] suffering can become an unmitigated scar, warping a personality with bitterness and resentment. But just as sunshine, which turns milk sour, will make apples sweet so the experience of suffering need not be evil in the end. We have a measure of free will and can choose what attitude to adopt to our fortunes however disastrous (a truth to which even the tormented inmates of concentration camps have been able to bear witness[9]). The mystery is that suffering borne willingly with love and, if need be, forgiveness, can become a source of goodness and life. 'Blessed is the man . . . who going through the vale of misery uses it for a well', wrote the inspired translator of Psalm 84:6.

The symbol of the cross offers a path we may follow through pain, disaster and misery and it leads to something worthwhile. Even what may have seemed evil can be transformed, but only those who follow this path (and many have) can know the resurrection that lies beyond the cross. It is a hard route and not tested easily; a consolation is that God travels the path with us and has already gone before us. This belief is at the heart of the Christian gospel.

I have put great emphasis on experience when discussing how religious beliefs may be tested. They can only be known to be true or false by living them; they have an empirical testable component. We should beware, however, when investigating the truth claims of faith, of making too much of the distinction between religious and non-religious experiences. For convenience it is frequently useful to be able to separate things into groups, labelling them appropriately. William James did this for the experiential dimension of religion when he wrote his great work *The Varieties of Religious Experience* at the turn of the century. Research along these lines has been conducted recently at Manchester College, Oxford where the biologist Dr. Alister Hardy has accumulated many thousands of accounts by ordinary people of experiences which fall into the 'religious' category. They were in answer to the question 'Have you ever been aware of or influenced by a presence or power, whether referred to as God or not, which is different from your everyday self?'[10] Nevertheless, despite this departmentalising tendency, we may still feel a reasonable urge to resist it on the grounds that the thoughts and feelings which arise out of our faith are better not boxed up in the word 'religious'. Can we really distinguish for instance – as I was once asked to in a questionnaire – between the aesthetic impact of a great cathedral and the religious impression it makes on us? There is no 'religious department' to life if the whole of life is an expression of the creative will of God, and perhaps we should be suspicious of any move to establish an apartheid between the sacred and the profane.

William James seems to have come to this conclusion towards the end of his life. In an essay 'A suggestion about mysticism' he argued that so-called mystical experience is not abnormal or supernormal but a perfectly ordinary extension of our normal field of consciousness. Consciousness can suddenly widen, the clouds lift as it were, and we see things more deeply and significantly; it is a sense of expanding reality, the sort of event described by the psychologist Abraham Maslow as a 'peak experience'.[11] Just as mountain peaks have their bases in the plains so 'religious experiences' are continuous with ordinary experiences however much they may stand out in memory as startling or transcendent.

In their own special way the articles of faith can indeed be tested. They have components which may be corroborated and which are potentially falsifiable, even though there are bound to be disagreements about what counts as evidence of falsification. There can never be the detached rigour of a classical scientific experiment because religious beliefs more than anything else involve the total interlocking of the observer with the statements being tested. All religious beliefs are in the end about people and their relationships with themselves, other people and their physical environment. They are hypotheses for living. The very idea of a detached observer is therefore a nonsense. It is only in the context of relationships that the true content of beliefs can be examined. It is in these areas that the symbol of God may begin to have meaning. There are of course unique problems here in assessing the value of the symbols of faith and the conclusions we reach are bound to be subjective but are *not thereby untrue*.

Our religious beliefs and value judgements are in the very nature of things subjective, they could not be otherwise. But they are subjective views of an objective reality. We may not like or even understand the subjective view of a landscape which an artist expresses on canvas but that does not lead us to deny the objective reality of the landscape itself. Religious statements about God, people, and the world are analogous.

Testing religious beliefs involves an 'inner game', an extension of a game we already play as we explore the way we relate to the world around us, to other people and to our own spirits. It is a process of reorientation and when dramatic may even be called conversion. For each of us this will mean finding our own inner story or myth. Our ancestors found meaning through myths and many people feel an atavistic nostalgia for the days when myths had power. But that dream-time of humanity is past and the old stories cannot be artificially revived; the Wagnerian method does not really work for religion. We live in a new world with a wealth of scientific knowledge. By quiet attentiveness and contemplation we make ourselves prone to inspiration. As we integrate the symbols of faith into our own lives we will each of us find our own myth, our own inner story, which will hold all things together in a coherent vision. This vision of meaning cannot be forced but like other organic things needs nurturing that it may grow naturally. I do not imply, however, that each of us will go it alone; we share all the great symbols in common. This is one universe evolving in one divine ground; spirit and matter are two aspects of one unfolding reality. In each of us individually, and in humankind corporately, God seeks his incarnate body.

There are no other tools we can use in these experiments with faith, but ourselves. Immersed in and emerging from the process of life on earth we in our wholeness are the instruments of measurement.

9. HUMANS AND THE ENVIRONMENT

The earth, viewed from space, is a beautiful bright ball of colour, a solitary sphere suspended by nothing in emptiness. According to Dave Scott, who landed at Hadley Rille with Apollo 15, the earth hangs motionless in the black lunar sky like a brilliant blue and white Christmas tree decoration. Dave Scott and his colleague Jim Irwin found that when they looked back at our world from their rare vantage point on the perimeter of the Sea of Rains they could 'hold' the earth between finger and thumb. Their novel perspective has given us a new symbol, something we knew but had never entirely grasped; we live in *one* world. Planet Earth, wrapped in its thin film of atmosphere, ocean and life (relatively no thicker than the bloom on a peach) is a single functioning entity. Its ecosystem is an interconnecting interdependent web of life; so much so it is almost as though it were one complex living organism.

It will take time for this vision to take hold of humankind's corporate mind. After NASA had extracted every last digit of information from the astronauts there was still something left untouched. Dave Scott told me once in conversation that he had been trained as a pilot and as a scientist but that had not prepared him for another aspect of the mission which he felt was deeply significant. 'The philosophical and the moral implications of that mission have never been drawn out' he said several times. It is as though a new mythology is being born alongside the development of science. Our conception of the world will never be the same again.

The moon in comparison is a dead world but very beautiful nevertheless. Any amateur astronomer with as little aid as a pair of binoculars can be wrapped in enchantment by our silent neighbour in space. My own fascination with the moon began when I was nine, viewing it through the branches of a beech tree with a pair of captured U-boat binoculars lent to my father. A small telescope will show clearly the ancient ruined craters many of them billions of years old and the sharp black shadows cast by solitary mountain peaks across the crystal clarity of sunlit seas of solidified lava. Even the names evoke mystery; the Bay of Rainbows, the Sea of Nectar, The Ocean of Storms or the Marsh of Decay; vast craters called Ptolemy, Aristarchus or Eratosthenes. But the mass of the moon is not great enough to retain an atmosphere and so it has never fostered a process of organic evolution despite the fact that it contains all the necessary elements for life. No woodlarks sing or owls hoot beneath a copper tinted blue night sky as they do by moonlight on earth. The mystery of life lies dormant in the rocks of the moon.

We on earth have emerged from, and are products of, the environment, and yet it is a peculiar deficiency of people today to feel alienated from nature. We live as though we were strangers in the land. Our attitude is that of careless colonisers pillaging someone else's property. Urban society is separated from the soil by a layer of concrete; the seasons of seed time and harvest have become irrelevant as transport and deep freeze technology provide food or flowers at all times of the year. Daffodils, new potatoes and freshly podded peas no longer herald the spring. We belong intimately to the earth and yet in our eyes it has become a resource to be exploited; its only value lies in its usefulness.

Ecological problems arising out of our careless relationship with nature have been forcing their way insistently to the front of our consciousness since the 1960s. Radioactive waste is polluting the seas while acid rain poisons our skies, rivers and forests. Bad farming turns good land into dust bowls, the destruction of rain forests changes the climate, insecticides kill the rich life of the hedgerows. Hanging over all these problems there is the ultimate threat of a deadly nuclear winter. A Christian vision of human meaning and purpose will remain flawed if it does not address itself to these issues. A theology of nature which encourages respect for, rather than a heartless rape of, the environment must become an essential ingredient in any modern framework of faith. The proliferation of pressure groups founded in response to the ecological threats that face the world is a healthy and hopeful sign that human beings will not remain a sleepwalking species forever.

The theology of nature has gone through a number of phases. Pre-Renaissance thinkers would have attributed our sense of estrangement from nature to the fall. God made the world to be paradise, but Adam and Eve were turned out of the garden of Eden when they fell from grace. The world became for them and their descendants a hard place full of toil, pain and sorrow. But a new respect for nature came with the Renaissance and writers as early as Francis Bacon in the seventeenth century limited the effects of the fall to the moral realm. Nature remained untainted by the corrupting effects of original sin, they argued, and continued to manifest the perfect work of its omnipotent creator. Like God in the Genesis myth they 'saw that it was good'.

The eighteenth-century deists built on this new perspective and argued that the perfect mechanisms of the universe demonstrated clearly the existence of God. Sir Thomas Browne wrote 'There are two books from whence I collect my Divinity, beside that one written by God, another of his servant Nature that universal and public manuscript that lies expans'd unto the Eye of all; those that never saw Him in the one, have discovered Him in the other.'[1] The progress of science was strengthened by this new vision; the curiosity of Renaissance thinkers was awakened and fascinated by the exquisitely detailed and beautiful order of things. God's artfulness was being revealed under the microscope and through the telescope.

This new urge to understand God's world, however, spawned its own estrangement. The scientific approach to nature has generated a new sense of our alienation from nature. The analysis of the observable universe into its constituent parts tends to reduce what is beautiful and inspiring to dead bits and pieces. Wordsworth expressed this fear in his famous lines:

> sweet is the love which Nature brings
> our meddling intellect
> Mis-shapes the beauteous form of things:–
> We murder to dissect.

Another writer who believed that our capacity for wonder was threatened by science was Goethe. He even, surprisingly, inveighed against repetitions of Newton's classic experiments with prisms with which he split sunlight up into the richly coloured bands of the spectrum!

> Friends avoid the darkened prisons
> where they pinch and tweak the light.

The anxiety expressed by these and many other writers was that by cutting nature up and reducing it to its parts we would become progressively disenchanted with the world and so would cease to value it. In exploring God's creation we might lose our consciousness of God.

Critics of the methods of science did not limit their warnings to the dangers inherent in reductionism. They also denounced any sacriligious attempts to interfere with or to change nature. Jean-Jacques Rousseau writing to an American gardening niece in the eighteenth century instructed her to have nothing to do with double primroses; God has never intended them to be like that, he pronounced. One of the biggest adjustments theology has had to make is to accept the idea that we live in a world where change is the norm. God did not make a static unchanging universe; God *is* making an evolving unfolding one; as conscious products of the process we have become instigators of change ourselves. By the twentieth century a Pandora's box of manipulation and interference had been let loose on the environment. Or will it, perhaps, be seen one day to be a cornucopia? Our meddling may be the creative response God expects of us. The double primrose bloomed through the cracks of a static model of the universe and we are just beginning to take seriously the thought that we inhabit a flowing process, one which we may begin to direct.

Mammoth ethical dilemmas now confront humankind. They cannot be sidestepped and we cannot put the clock back and pretend they do not exist. It was inevitable that our exploring spirit would reveal the means for selective breeding and for genetic engineering. It is a consequence of our understanding the climate that we may be able to manipulate the weather. It was unavoidable that science would discover the secrets of the atom and find ways for unlocking its fabulous energy. If we are to face all these things with responsibility and wisdom then a theology of nature becomes an essential aspect of our faith. It will take a bold attitude to change and will help us to see our relationship with nature in creative and loving terms.

Christian theology on the whole has been deficient in its attitude to the material world. Two realms have traditionally been its major concern; the relationship between the individual soul and God and the relationship between the individual and society. Love of God and love of neighbour threw nature into the shade. Apart from some devotional literature like St. Francis' hymn to brother sun,[2] a theology of nature has never been adequately developed, certainly not one which can give guidance to twentieth-century people facing entirely novel moral dilemmas. The purpose of such a theology will be to proclaim redemption, to produce a reconciliation between humans and the environment, from which they feel alienated. Through selfishness and half wakefulness our shortsighted view of this life has caused us to become estranged from the world we are born into. A new vision is needed to recreate a sense of wholeness and to establish the intrinsic worth of the natural world.

Science and theology together reveal that all the abundant richness of reality is rooted in an underlying unity. The Christian view of the universe is that everything is the product of the will of one God, 'maker of heaven and earth and of all things visible and invisible'. The insight of physics is that the immense diversity of nature, from galaxies of stars to the complex chemistry of butterflies and human brains, is all the result of the interactions of a few basic types of particle with three types of force. There is a deep and simple unifying coherence about the world we inhabit.

Science goes further and tells us something about the fundamental symbiosis of all living things on earth. Life on earth ultimately has to be seen as one individual phenomenon. It is a single great enterprise of which we are a tiny dependent but contributory part.

Science is a two-way process. It probes for detail, analyses, dissects and examines but it also has another mode which looks for the connections between things. One modern writer suggests, 'The question then is whether there is not a built in drive in the logic of explanation that impels it to seek the connection that connects all the connections.'[3] Religious faith and science both seek for the deep down unity behind surface reality. This urge has led to an interdisciplinary approach resulting in new subjects such as biochemistry or sociobiology for instance. It has also led among other things to the concept of Gaia.

Gaia or Ge, the earth, from which we get words like geology and geography was the first being, a goddess, according to Greek tradition, to be born from the primeval chaos. The Gaia hypothesis treats life on earth as a single though complex evolving phenomenon. The millions of types of organisms that dwell on the surface of the planet, plants, insects, fish, birds, mammals and so forth are not a random collection of discrete systems of evolution. They form an interdependent community of evolving activity. The ecosystem which wraps itself around our planet has been compared to a single cell in which all the parts have their own individual functions. As life evolved it slowly transformed the environment to suit its own interests. Evolution fuelled its own development by breathing oxygen into the atmosphere by photosynthesis; it grew its own protective 'membrane' of ozone thirty miles up which filters out the lethal ultraviolet rays from the sun.[4] Mother Earth, Gaia, functions as a self-correcting unit, an evolving enterprise of breathtaking complexity. From within the living ecosystem of this planet-sized cell-like process there have emerged creatures made in the image of God. Our sense of alienation from nature is a delusion, a product of ignorance and blinkered vision — an earlier theology might call it sin. We need the living environment that spreads across the face of the world beneath the protective sky even more than we need our arms and legs.

Evidence that life on earth is a commonwealth of complex connections is not hard to come by. Events in one part of the world have repercussions in other places far removed. Hints of this interconnectedness of things occur all the time. The famine in Africa of the past decade has been an immense human tragedy and for the first time in history humanity has been able to make a global, if inadequate, response. But a smaller detail escaped many people's notice though it illustrates how life across the earth is like a web; touch one part and the rest vibrates. In 1984 birdwatchers noted that not so many swallows came with the spring and fewer whitethroats were found nesting in English hedgerows. The reason was that the drought which had devastated the Sahara in Africa had enlarged the area of desert and many small migratory birds could no longer make the crossing. They perished in the sand. The trailing wings of other people's famines drag round the world.

The ease with which local ecosystems may be disrupted and changed has been demonstrated many times. An early enthusiasm for DDT produced unintentional chain reactions, sometimes with disastrous results. Some years ago the World Health Organisation attempted to eliminate the malaria carrying mosquito in Borneo. Unfortunately the DDT also killed a type of predatory wasp which normally kept a population of caterpillars in check. The ravenous caterpillars then ate the roofs of houses which began to collapse. Indoors, DDT was used to kill flies, the flies were eaten by gecko lizards and the lizards by cats. The cats died and there was a population explosion amongst the local rats; with them came the threat of plague. New cats had to be parachuted into the region to restore some balance in the local ecosystem.

The proliferation of pressure groups and the interest of the media in anything that smells of disaster is awakening us all to the importance of understanding the dynamic equilibriums that prevail both locally and globally. The issues even filter through to the teaching syllabuses of schools. The consequences of polluting the atmosphere with acid rain from the burning of fossil fuels; of cutting down the tropical rain forests; of overfishing the seas; of turning prairies into dust bowls through bad agricultural husbandry, list but a few of the concerns arising from the vibrations that humankind sends yearly through the ecosystem of the planet. Economics, market pressures and political expediency will not be enough to regulate these things wisely. Only a thoroughgoing coherent faith with a vision of the human role in the world can provide us with the will and the motivation to get it right. Ecology needs theology.

The guiding concept for the Christian faith will be the doctrine of the incarnation. The belief that in the person of Jesus Christ God became a man has radical consequences for our view of the material world. The material universe is not an evil realm from which the spirit must escape. Spirit and matter are not two separate realities but are aspects of one reality, a spirit-matter unity that is rooted in God. By the incarnation God, the ground of the universe, became flesh and revealed that chemistry may become a vehicle for divinity. A fundamental insight that emerges is that matter matters; the wholehearted acceptance and high valuation of the material substance of the world is a distinguishing feature of Christian spirituality.

All flesh, lovable and vulnerable, tender and cancer-prone must be seen from the point of view of the incarnation. Flesh itself is made from the dust of the earth and so from atoms which ultimately were forged in the cosmic furnaces of giant stars aeons ago. The same atoms have been recycled numberless times on the earth's surface. In the incarnation God is manifest in the flesh and so all matter glows with a new significance. The deep link between spirit and matter is revealed. The distinction between the sacred and the profane, the holy and the non-religious, no longer applies except in the drama of religious liturgy. For those who see the world in terms of spirit-matter and of the incarnation, nothing will appear profane; our physical environment, though often plaguing us with catastrophes and pain and from which people today are in danger of feeling alienated, is an aspect of sacred reality.

Two levels of description are needed to give an adequate account of the person Jesus Christ. At one level he was a member of the human race, a product of DNA inherited from his mother (and father? This question which we considered earlier must remain open). Physically he was Jewish and claimed descent from King David. His chemistry was like that of other mammals; he shared in the gene pool of humankind and the genes we have in common with other creatures he also possessed. He lived by eating; his body was sustained and daily renewed by consuming elements from the soil transformed into plants and animals. All the atoms of his body, with those from which we are built, had drifted through space for billions of years before the solar system came to be and the evolution of life began its long process causing mind to emerge from matter. At another level Jesus Christ is the Logos, the word of God. In him the divine ground of the universe and the biochemistry of evolution become one. This is the mystery of the incarnation and this is why the material universe must have such a high place in a Christian theology.

The deep unity between God and matter is not limited to the flesh and blood of Jesus Christ. It is a potential for all of creation. When Jesus took bread and broke it with the words 'this is my body' and over the cup of wine said 'this is my blood', he was indicating a new way of looking at the ordinary elements of an ordinary meal. The living presence of Christ in the eucharist is already known and experienced in a great variety of ways in the worldwide church; from the liturgical formality of a Latin mass to the plain informality of a Methodist house church. It is known as much in a Quaker silence as it is in the choral glory of a cathedral celebration of the mysteries.

The material universe takes on a new meaning when seen in sacramental terms. There is more to reality than what is outward and visible. The inward and invisible is as real as anything that can be described by science. Christ seeks a body in this material universe, a process which Teilhard de Chardin called Christogenesis. This movement towards incarnation is an expression of the dynamic activity of God which Christianity has attempted to formulate in the doctrine of the trinity.

Teilhard de Chardin contemplating the sacrificial language and imagery of Jesus' crucifixion saw in it a way of understanding evolution.[5] The pain and suffering, the toil and tears of human life and of the whole evolutionary process were illuminated for him by his own breaking of bread as a priest celebrating the mass. As he pursued his palaeontological interests in humankind's origin, travelling in remote parts of Asia, he was often the only Catholic for hundreds of miles. Unable to celebrate the eucharist with a community he turned it into a mystical contemplation of nature. The altar was the world and the priestly action proclaiming 'This is my Body' over the broken bread extended beyond the transubstantiated host to the cosmos itself. The whole world is to experience incarnation as it transforms itself into God who animates it. Incarnation is not only unfinished, it has only just begun.

The thought that Christ might be found as a living presence elsewhere than in the communion meal is not new. An earlier writer of New Testament times had a similar mystical vision of the presence of Christ in creation. A collection of papyri found at Oxyrhynchus in Egypt contained a fragment from the third century of our era. It has been identified as part of 'The Gospel according to Thomas' and it records an otherwise unknown saying attributed to Jesus:

> Turn the stone and
> I am there
> Split the wood and you will find me.[6]

The Christian faith must be ready to look for the incarnation not only in the historical event of Jesus' life or in the sacrament of holy communion; Christ is to be sought in other people, in marriage, in families, in schools, churches and all other institutions. Wherever there is community Christ seeks a body. The passion narrative of the last days of Jesus' life are a parable of the rejections, divisions and crucifixion that happen whenever people resist the embodiment of God in their world. Even the universe itself is groaning for the day of a cosmic incarnation (Rom 8:22), a resurrection from death and ignorance to life and consciousness of God.

The link between the incarnation and the rest of creation was established in fact in the New Testament but it is only recently that the implications for ecology are being drawn out. A fundamental question asked by children and theologians is 'Why did God make the world?' The New Testament answer involves the belief that the word, the Logos, which was made flesh was also the instrument by which God created all things. Early church fathers interpreting this doctrine had to steer a careful path between two heresies which put limits on the perfection of God. One implied that God created the world because God felt incomplete without a universe to love while the other suggested that God had no choice in the matter because it was part of his nature to create. Their language was often tortuous and difficult to grasp but in the end they were quite clear that the cosmos is an expression of God's free rational will. It is the result of the creative activity of his Logos. The same mind which gave logical (the same root as *logos*) form to the universe is that which we find incarnate as flesh and blood in Palestine. The creation of the world and the incarnation of Jesus Christ are both the result of God's creative word, 'Let there be ight'.

New Testament writers, meditating on the cosmic significance of Christ drew liberally on the wisdom literature of the Old Testament and the Apocrypha. The Judaic notion of wisdom personified as a woman and the Greek idea of the Logos flow together and illuminate one another in this contemplation of the meaning of the incarnation. Christ is described as 'the first-born of all creation' (Col 1:15, JB) and 'the effulgence of God's splendour' (Heb 1:3, NEB), phrases which borrow heavily from the imagery of Wisdom 7:25.

People today must learn to coexist with nature, not conquer it. A new theology of hope proclaiming the meaning of human life in the context of nature will grow naturally from the fundamentals of Christian revelation. Partial truth (which is the meaning of heresy) has dominated the Judaeo-Christian attitude to the natural world and is only beginning to be put right. Humankind stands on the brink of the most dangerous era in its career; knowledge without wisdom could spell the end of the species. The creating Spirit of the universe invites men and women to become co-creators; the incarnation of the word of God must become a reality in our own lives in the twentieth century so that we may face the future with rationality and wisdom.

The partial truth which has held sway over the Judaeo-Christian mind is the aggressive heretical theology of domination. The natural world is held to be inferior to humanity and so must be dominated, despite the opinion voiced in the Genesis myth that at each stage of creation God 'saw that it was good'. The material and living environment was never allowed to be of value in its own right. After the legend of the great flood humans are established quite clearly as lords of Earth: 'And the fear of you and the dread of you shall be upon every beast of the earth . . .' (Gen 9:2, AV). This theology of domination gave licence for the misuse and neglect of nature. The balance needs to be redressed. Humankind's dominion over the world is misinterpreted if it is taken to mean that it owns it as a medieval Russian landlord owned his serfs. A Christian understanding of lordship is that it implies service and is illuminated by the image of Jesus washing the feet of his disciples. In the spirit of St. Francis, the scientific person should perceive an intimate kinship and care between herself and the world so that we may speak of brother sun or sister moon, brother atom or sister chromosome.

The overriding preoccupation of scripture is with God, people and society; very little is said of the environment. Apart from the Levitical law to give the land a sabbath rest every seventh year and a command not to cut down the fruit trees when besieging a city (Deut 20:19) – a practical and touching detail following a verse in which the Israelites were instructed to practise genocide against various tribes who frustrated their plan to colonise the land! – there are very few references to nature.

A whole new area of ethics must be developed to cope with humankind's increasing control of its environment. There is a rational component in ethics which faith may develop and we must widen the circle of our compassionate concern. Sociobiologists speculate that people inherit various behavioural traits from their ancestors. Humans are social and they have a natural inclination transmitted by DNA to live cooperatively with close relatives. Altruistic behaviour within your own group increases the chance of the group's survival. Caring and loving may originally have been part of our biochemical make-up. Along with this inward-looking helpfulness of the group (a sort of behavioural 'charity begins at home') goes an equally strong disregard for 'them' who do not belong to the group. Both Jesus (Mt 5:43) and Plato[7] referred to the prevalence of the moral opinion that it was quite proper to love your friends and hate or harm your enemies; both advocated widening the circle of love and justice. Imagination inspired by compassion and guided by reason can extend the frontiers of moral responsibility. Maybe we are now on the verge of realising for the first time in history that we are members of a global village. The environment for which we feel a growing sense of kinship must include people as well as fruit trees. Racism should have no place in a world inhabited by *one* human race all mythically descended from the same ancestors Adam and Eve.[8] It is not enough in this waking world for us to do what is natural and to believe that what we feel in our bones is best. The treatment of any societies as second class or of nature as inferior must be impossible within a Christian vision of the world. The vocation to be human demands that we transcend the prejudices dictated by DNA and develop a new deeper more God-like way of being persons.

The discovery by science of nuclear power is perhaps the greatest cause of disquiet today. The threat of a nuclear holocaust laying waste to the earth, the result of a moment's irrational madness or, even more sickeningly, a mistake, is the most horrific scenario we can contemplate. Even as a source of energy for peaceful purposes it raises justifiable fears. The dangers of radiation are now abundantly clear; there are problems of transporting nuclear waste products, of their disposal and storage, and of their use legally or illegally for making nuclear bombs. Nuclear reactors which have had mishaps and are infilled with concrete will stand, we are told, for thousands of years before they become safe; the tombstones of an overhastily developed technological civilisation, some fear. The misuse of pesticides contemplated by Rachel Carson in her book *Silent Spring* is as nothing compared to the dangers resulting from the misuse of nuclear knowledge.

The clock cannot be set back however. Nuclear knowledge is here to stay: the question is *how* we use it not *whether* we use it. August 6th will forever remain a memorial day of human misuse of technology; in 1945 150,000 people were incinerated by the grotesque nuclear blockbuster euphemistically named 'Little Boy', and thousands more suffered the horrendous aftereffects of radiation. But from that twentieth-century crucifixion good may come if we will it. If the world has learnt a lesson from Hiroshima (and let us not forget Nagasaki) then those Japanese cities may have saved us from much worse.[9]

Only one gram of matter, a featherweight, transformed into pure energy in accord with Einstein's formula $E = mc^2$ (E = Energy, m = mass, c = speed of light) was enough to lay waste to a whole city; such is the stupendous energy locked within the atom. We cannot set this knowledge on one side as was naively suggested by an Anglican bishop in a letter to *The Times* not long after World War II. He proposed that the allies tear up the formula as though it was rather like the secret for making Benedictine which once forgotten would be lost forever. Scientific knowledge is not like that. The laws of physics are a coherent system of which our understanding of the mechanisms of nuclear fission (basic to the atomic bomb in the splitting of uranium atoms) and nuclear fusion (basic to the hydrogen bomb in the combining of hydrogen atoms to form helium) are but a small part. It would be quite impossible both practically and intellectually for the scientific community to censor this particular bit of knowledge.

Christian faith must be bold and not cower from nuclear energy as though it were an evil genie let out of a bottle by immoral scientists. God has built this universe of stars and people on the principle of nuclear power and has given us minds to comprehend it. There is great beauty to be found in contemplating the immense energy resting at the heart of all matter. The suns that created the atoms from which we are built ran according to the dictates of atomic law; they burned by nuclear fusion. Our own sun which providentially gives us warmth and light is in a continuous state of nuclear explosion. It remains in equilibrium because the gravitational attraction of its great mass exactly balances the outward explosive force of the nuclear energy deep in its interior where the temperature is estimated to be above twenty million degrees K. The whole evolving ecosystem of this planet from plankton to people depends for its life upon the same principle that gave rise to the nuclear warheads in Trident or SS100 missiles.

Isaiah's vision of swords being turned into ploughshares is as relevant as it ever was. There are many peaceful purposes to which knowledge of nuclear matters can be put. Our approach to the use of this scientific expertise should be positive and optimistic and not merely a reluctant acquiescence with a malignant technology. Experiments at the Culham laboratories near Oxford and elsewhere point the way to the possible use of nuclear fusion for peaceful purposes – to harness the power of the atom for the national grid. There are far less dangers involved in extracting energy this way than in the present conventional method using uranium and the nuclear fission method. Fossil fuels will eventually run out however economically we use them. If people have the great future we hope for, then nuclear energy is bound to become one of its realistic options; for generating electricity, for powering ocean-going liners and space ships; for large engineering projects that involve shifting large quantities of rock; or as we have seen in an earlier chapter, for controlling the doomsday flight path of any future asteroid destined to collide with earth. With vision and commitment to the belief that this universe is the result of a deliberate act of creation and is not a meaningless accident, humankind will easily overcome the problems posed by the scientific research which has exposed the energy locked within the atom. The nuclear winter need be no more than the bad dream of a civilisation still in its adolescence.

Other major issues raised by science will have to be approached with the same mixture of knowledge, vision and courage. We are treading new ground and need perforce to create new values. The problems we face will not go away. We live on a small finite planet and our population is expanding. We depend on fossil fuels which will run out at least in the days of our great great grandchildren. The agricultural resources of our world are limited. No parent who knows it is possible to avoid giving birth to a handicapped child will refuse the benefits of embryo research. In the future mammoth unpredicted disasters may threaten our distant descendants; we cannot deprive them of the knowledge and expertise which may protect them from harm.

Already we can see some of the positive fruits of the exercise of conscious control over nature. Smallpox was eradicated in 1980, eliminated for ever by a campaign conducted by the World Health Authority. (Will the malaria virus which kills almost a million people per year be next to be conquered?) We crossbreed cattle for leaner meat and manipulate the genes in wheat to produce new heavy cropping strains; insulin is manufactured using bacteria; contraceptive methods take some of the haphazardness out of childbirth. Every time we build a dam or plant a forest we change the environment. We even allow the feeding of poisoned sandwiches, by a bird protection society, to common gulls (was this the right ethical choice?) to limit their numbers so that the rare and beautiful roseate terns have a better chance of survival. There are no moral absolutes in any of these greater or lesser issues but our response to them must be creative, loving and responsible. There is great danger here posed by a sort of moral fundamentalism and a temptation to cling to absolutes as though that were a mark of Christian integrity. We have to believe in the good and the beautiful without always being certain of where it lies. Many of the moral dilemmas that face humanity involve big risks whichever way we decide to turn. But as creatures made in the image of God we have the authority to create values in novel circumstances. Often it will be the way we live with, and see through, the moral decisions we make which will be more significant than the actual decisions themselves. Prescientific attitudes to the world enshrined in rural arcadian dreams will never solve the real problems that now concern twentieth-century people. There is no way back.

The word paradise comes from a Persian word for a park; it has become a symbol of a living order and beauty where value and fact are undivided. When a Christian uses the Lord's prayer and prays 'Thy kingdom come. Thy will be done in earth as it is in heaven' (Mt 6:10, AV) he or she should not just have in mind the city of God as a community of people. A physical paradise can be part of the vision. The material and spiritual continuity between human beings and the rest of evolving life has only just dawned on us. Not only are we ascended from the primate branch of the mammalian group of vertebrates but we are even cousins to the wheat of the field from which we bake our daily bread; ancestral enzymes in each of us proves the point. The interplay of law and chance has woven both the wild scabious in the August hedgerow and the brain of the nuclear physicist. As products of evolution and conscious self-aware creatures we must recognise our dual role as lords and servants of nature; we are bound to honour and respect our parent the ecosystem which has helped give us birth.

Responsibility needs to be tough. Respect for nature and a

proper concern for the ecological problems that face us will not mean that we have to adopt a sentimental 'preserve everything' environmentalism. Science, the greatest tool of the western mind (and technology its practical application) is itself an emergent phenomenon of evolution. We will have to be bold in our use of it and inevitably take risks; the dilemmas we face will rarely present easy options. But we plan and create in an evolving environment which itself is rooted in the divine ground; 'in him we live and move and have our being' (Acts 17:28, NIV). Spirit and matter are two aspects of a deeper reality.

God is at the heart of this value-creating meaning-seeking process that we inhabit; the cosmos is not the result of an impersonal haphazard accident. Change is as normal as static harmony in this world and sometimes we may have to let go what we would like to preserve; to take risks where we would prefer to seek security. We will find beauty as much in the urban concrete landscapes as we do in the rural. When the clouds hide the top of the Pan Am building on Park Avenue and the steam escapes from the ducts beneath the noisy Manhattan pavements a New Yorker can be as thrilled with the wonder of the place every bit as much as the poet wandering through the lakeland hills. We are not strangers in the land. We are the most wide awake of all the products of evolution, visionary and farsighted, with power to change the world for good or ill. 'In the beginning . . .' is *now*.

NOTES

CHAPTER ONE

1. *Simone Weil*, **The Need For Roots**, trans. A. F. Wills (Routledge and Kegan Paul 1952) 250
2. *Lesslie Newbigin*, **The other side of 1984**, The Risk Book Series (World Council of Churches 1984)
3. *Paul Davies*, 'The Eleven Dimensions of Reality' **New Scientist** 9 February 1984
4. *Lewis Thomas*, **The Medusa and The Snail: Notes of a biology watcher**, (Penguin 1981)
5. *St. Athanasius*, **On the Incarnation (De Incarnatione)** (Mowbray 1953) section 54

Further Reading

Peter Medawar, **The Limits of Science** (OUP 1985)

Frank Close, **The Cosmic Onion: Quarks and The Nature of the Universe** (Heinemann Educ. Books 1983)

Paul Davies, **God and The New Physics** (J. M. Dent and Sons Ltd 1983) ch. 8 'The quantum factor' 100 ff

Russell Stannard, **Science and the Renewal of Belief**, (SCM 1982) ch. 18 The Place of Paradox in Science & Belief

John Stewart Collis, **Vision of Glory** (Penguin 1972) 'The new alchemy'

Fritjof Capra, **The Tao of Physics. An exploration of the parallels between modern physics and eastern mysticism** (Fontana/Collins 1975)

Gary Zukav, **The Dancing Wu Li Masters. An overview of the New Physics** (William Morrow and Co. NY 1979)

Ursula King, 'Modern Cosmology and Eastern Thought, Especially Hinduism' in David Tracy, Nicholas Lash (eds.) **Concilium:**

Cosmology and Theology (T. and T. Clark/The Seabury Press
1983) 76–83

A. *Koestler*, The Sleepwalkers, (Hutchinson 1959)

Martin Werner, The Formation of Christian Dogma, (Adam and
Charles Black 1957) p. 196 Chapter on Christ's work of redemp-
tion Bk. 2. Section 2: The formation of the New Doctrine of
Christ's Work of Redemption p. 193, ch. 1 The Soleriological
Significance of the Incarnation of the Logos. pp. 199–211, ch. 2
The Sacramental Significance of the Death of Jesus.

A. *Koestler and J. R. Smythes*, Beyond Reductionism: New Perspectives in the Life Science. Proceedings of the Albach Symposium.
(Hutchinson 1968 [out of print])

CHAPTER TWO

1. *J. Klepesta, A. Rükl*, Constellations. A Concise Guide in
   Colour (Hamlyn 1969)
2. Quoted by Jeremy Cherfas in New Scientist, 17 May 1984
3. Plato, Timaeus and Critias, trans. Desmond Lee (Penguin
   Classic 1965) p. 123
4. *John Baptist de Freval*, Nature Displayed (London 1735)
5. *Fred Hoyle, N. C. Wickramasinghe*, Lifecloud: The Origin of
   Life in the Universe (Dent and Sons Ltd 1978)
6. *F. Vasek*, Am. J. Bot 67 246–255 (1980)

Further Reading

A. P. *Norton*, Norton's Star Atlas (Gall and Inglis 1973)

Adam Ford, 'Stargazers Guide to the Night Sky', a series of audio
cassettes, available from Terra Firma cassettes, 55 Bolingbroke
Road, London W14 0AH

D. *Malin, P. Murdin*, Colours of the Stars (CUP 1984)

CHAPTER THREE

1. *S. Weinberg*, The First Three Minutes (A. Deutsch Ltd.
   London 1977)

2. *Koestler (ed)*, **Beyond Reductionism**
3. *Peter Atkins*, **The Creation** (W. H. Freeman 1981) 126
4. *Brandon Carter*, 'Large number coincidences and the anthropic principle in cosmology', in M. S. Longair (ed.) **Confrontation of cosmological theories with observation** (Reidel, Dordrecht 1974)
5. *Richard Feynman*, **The character of physical law** (BBC Publications 1965), 125
6. *Karl Popper*, **The Open Society and its enemies, I Plato** (Routledge and Kegan Paul 1945) 12
7. *Kenneth Clark*, **Civilisation** (BBC Publications 1969) 129
8. *Peter Atkins*, **The Second Law** (Scientific American Library 1984) 200
9. *D. A. Mackenzie*, **Teutonic Myth and Legend** (Gresham Publishing Co)
10. *Stephen Powers*, **Tribes of California** (1877) (Univ. of California Press Republication 1976) 250, 273
11. *Hans Küng*, **Does God Exist?** (Collins 1980) 529
12. *Stephen Hawkins*, 'The Edge of space-time', **New Scientist** 16 August 1984, 10
13. *A. Flew, A. Macintyre (eds)*, **New Essays in Philosophical Theology**, Ch. 6 'Theology and Falsification' (SCM 1955) 96
14. **Masters of Prayer: Julian of Norwich**, Foreword by Pamela Searle (CIO publishing 1984) 20

Further Reading

B. W. Anderson, **The Living World of the Old Testament** (Longman 1979)
Paul Davies, **God and the New Physics** (Dent and Son 1983)
Paul Davies, **The Accidental Universe**, ch. 'The anthropic principle' (CUP 1982) pp. 110ff

CHAPTER FOUR

1. Asa Gray quote p. 7 B. Quartitch Ltd., Natural History Catalogue 986
2. *Michael Ruse*, **Darwin Up to Date: A New Scientist Guide** (1983)

'Creation Science: The Ultimate Fraud' in Jeremy Cherfas (ed.)

3. *E. Gilson*, **The Christian Philosophy of St. Augustine** (Gollancz 1961) 197–209

4. *H. B. D. Kettlewell*, 'Evolution and the Environment' in **Darwin Up to Date** 21

5. Museum of Natural History, South Kensington, Display about Evolution

6. *Stephen Jay Gould*, **The Panda's Thumb** (W. W. Norton and Co. New York 1980) 19–26

7. *William Paley*, **Natural Theology or Evidences of the Existence and Attributes of the Deity collected from the Appearances of Nature**, (1802)

8. Article 'on being religious', Posthumous Papers of D. H. Lawrence (Phoenix, Heinemann New York 1936)

9. *Richard Swinburne*, **The Existence of God** (Clarendon Press 1979) 133ff ·

10. *H. N. V. Temperley*, **A Scientist who believes in God** (Hodder and Stoughton 1961)

11. *H. N. V. Temperley*, 'Could Life have happened by accident?', **New Scientist** 19 August 1982 SOS

12. *Küng*, **Does God Exist?**, p. 645, Quotes Manfred Eigen **Das Spiel** ('The Game')

13. *D. T. Gish*, **Evolution: The fossils say no!**

14. *Jeremy Cherfas (ed.)*, **Darwin Up to Date** 8

15. **Epic of Gilgamesh**, Eng. version by N. K. Sandars (Penguin books 1960) 108

16. *J. Whitcomb and H. Morris*, **The Genesis Flood** (Presby and Reformed. 1960 [American Publication still in print])

17. *John Little*, 'Evolution: Myth, Metaphysics or Science?' **New Scientist**, 4 September 1980, 708

18. *Gould*, **The Panda's Thumb**, 'The Episodic Nature of Evolutionary Change' 179

19. *Mark Ridley*, 'Who doubts evolution?' in **Darwin Up to Date** 5–7

20. *Leonard Hodgson*, **For Faith and Freedom. The Gifford Lectures 1955–1957**, 2 vols (Blackwell Oxford 1957)

21. *Teilhard de Chardin*, **The Phenomenon of Man** (Collins 1959)

22. *Küng*, **Does God Exist?**, 347

23. *S. Aurobindo Ghose*, **The Life Divine** (India Library Society, New York 1949)

24. *S. Radhakrishnan Taittiriya Upanishad* in **The Principal Upanishads**, (George Allen and Unwin, London 1953) 562
25. *Teilhard de Chardin*, **Hymn of the Universe**, ch. 'The Spiritual Power of Matter' (Collins 1965) 65

Further Reading

Wilma George, **Darwin** (Modern Masters series) (Fontana 1982)
Jonathan Howard, **Darwin** (Past Masters series) (OUP 1982)
J. L. Mackie, **The Miracle of Theism**, ch. 8 'Arguments for Design' 133–149, (Clarendon Press 1982) 133H
Bryan Magee, **Popper** (Modern Masters series) (Fontana 1973)
D. T. Gish, **Evolution: The Fossils say No!** (3rd Edition 1973 Mark Books [American Publication still in print])

CHAPTER FIVE

1. 'Outer space: Is anybody there?' in **The Economist**, 7 March 1981
2. St. Athanasius, **On the Incarnation**, section 54
3. *R. L. Travers*, 'Foreword' in Richard Dawkins **The Selfish Gene** (Granada 1978) vii
4. *Jonathan Howard*, **Darwin**, Past Masters Series (OUP 1982) ch. 6
5. *Gould*, **The Panda's Thumb**, 'Natural Selection and the Brain: Darwin v. Wallace' 53ff
6. *ibid.*
7. *ibid.*, ch. 'The Episodic Nature of Evolutionary Change'
8. **Is Man a machine?** (Humanities: A foundation course. Unit 32) (Open University Press 1971)
9. *Igor Aleksander, Piers Burnett*, **Reinventing Man** (Kogan Page London 1983)
10. *John Hick*, **Evil and the God of Love**, Part 3: 'The Irenaean Type of Theodicy' (Macmillan 1966) 207–276
11. **Plato, Timaeus and Critias**, 121
12. *Tennyson*, **In Memoriam** (1850)
13. *Nicholas Berdyaev*, quoted by Mark Doughty, **The Tablet** 4 February 1984
14. *Colin Wilson*, **New Pathways in Psychology** (Gollancz 1962)

15. Apocryphal New Testament, New Temple Press 1820 – e.g.
p. 38. The First Gospel of the Infancy of Jesus Christ
16. *W. D. O'Flaherty*, **Hindu Myths**, (Penguin Classics 1975) 218
17. **Koran**, Sura 19: Mary
18. *H. V. Ditfurth*, **The Origins of Life**, trans. P. Heinigg (Harper and Row 1982) ch. 'The History of Cytochrome C'
19. *John Hick (ed.)*, **The Myth of God Incarnate** (SCM 1977)
20. *Peter Singer*, **The Expanding Circle Ethics and Sociobiology**, (OUP 1981)

Further Reading

Wagoner and F. E. Goodson, 'Does the mind matter?' in M. H. Marx, F. E. Goodson **Theories of Contemporary Psychology**, (Macmillan NY 1976) 201

CHAPTER SIX

1. Evolution Life: Nature Library p. 14
2. *David Day*, **The Doomsday Book of Animals** (Ebury Press 1981) 275
3. *Thomas*, **The Medusa and the Snail**, 107
4. News item, **New Scientist** 8 Dec. 1983 p. 719, Report on IRAS
5. *Natalie Angier*, 'Did Comets Kill Dinosaurs?' **Time** 6 May 1985 41
6. *T. S. Eliot*, **Four Quartets**, 'Burnt Norton' line 42 (Faber and Faber)
7. *Powers*, **Tribes of California**, 70
8. **Epic of Gilgamesh**, 108
9. *John Hick*, **Philosophy of Religion** ch. 4 'The Problem of Evil' (Prentice Hall 1983) 40–56
10. *Ninian Smart*, **Philosophers and Religious Truth**, ch. 6 'F. R. Tennant and the Problem of Evil' (SCM 1969) 139

Further Reading

M. Allaby, J. Lovelock, **The Great Extinction. What killed the dinosaurs and devasted the Earth?** (Secker and Warburg 1983)

Simon Miton (ed.), **The Cambridge Encyclopaedia of Astronomy**, (Jonathan Cape 1977)

John Hick, **Evil and the God of Love**

Teilhard de Chardin, **Hymn of the Universe**, ch. 'The Mass on the World' (Collins 1965)

Jim Garrison, **The darkness of God: Theology after Hiroshima** (SCM 1982) 105

CHAPTER SEVEN

1. *Stannard*, **Science and the Renewal of Belief**, Augustine quote p. 11
2. *Peter Medawar*, **The Limits of Science**, 41
3. *E. Pagels*, **The Gnostic Gospels**, (Weidenfeld and Nicolson 1980) 3

Further Reading

Smart, **Philosophers and Religious Truth** ch. 2 'Miracles and David Hume'

Brian Davies, **An introduction to the Philosophy of Religion**, ch. 11 Miracle (OUP 1982) 106–118

Geza Vermes, **Jesus the Jew** 'Excursus: *Son of God* and virgin birth' (Collins 1973/Fontana 1976) 213–222

CHAPTER EIGHT

1. *Magee*, **Popper**, 38
2. *Prof. Alan Smithers*, 'Ex astris ad infinitum' **The Guardian**, 22 March 1984, An analysis of 'The office of Population consensus and survey'
3. *T. S. Kuhn*, **The Structure of Scientific Revolutions** (Chicago 1970)
4. *Don Cupitt*, **Taking Leave of God** (SCM 1980)
5. *A. J. Arberry*, **Sufism. An Account of the Mystics of Islam** (George Allen and Unwin 1950, 1979 pbk) 80
6. *Eliot*, **The Four Quartets.** 'Burnt Norton' line 62

7. *Medawar*, **The Limits of Science,** 50
8. *T. S. Eliot*, **Murder in the Cathedral,** 50 The Four Tempters Speaking (Faber and Faber 1965)
9. *Viktor Frankl*, **Man's Search for Meaning** (Hodder and Stoughton 1964) 65
10. *Alister Hardy*, **The Spiritual Nature of Man** (Clarendon Press 1979) 18
11 *Colin Wilson*, **Maslow and the Post Freudian Revolution** (Gollancz 1972) 82/83

Further Reading

M. Gauquelin, **The Truth about Astrology** (Blackwell 1983 Hutchinson 1984)

H. J. Eysenck and D. K. B. Nias, **Astrology: Science or Superstition** (Penguin 1984)

Sir Charles Eliot, **Japanese Buddhism** (Routledge and Kegan Paul 1969) 197

CHAPTER NINE

1. *George S. Hendry*, **Theology of Nature** (The Westminster Press 1980) 56
2. **Francis and Clare: The Complete Works** (Trans. R. J. Armstrong and I. Brady) (SPCK 1982)
3. *Hendry*, **Theology of Nature,** 102
4. *Lewis Thomas*, **The Lives of a Cell** (Viking Press 1974), ch. 'The World's Biggest Membrane'
5. *C. F. Mooney*, **Teilhard de Chardin and the Mystery of Christ,** ch. 3 'The Incarnation and the Eucharist' (Collins 1966) 67–103 *Teilhard de Chardin*, **Hymn of the Universe** (Collins 1965), Chapter 'The Mass of the World'
6. *R. M. Grant, D. N. Freedman*, **The Secret Sayings of Jesus** (Fontana 1960) 45
7. *Plato*, **The Republic** (Trans. Desmond Lee) Book 1 v. 336, p. 73 (Penguin 1974) Discussion with Polemarchus
8. *K. Leech*, **Theology and Racism** (Board for Social Responsibility of The Church of England 1985)
9. *Jim Garrison*, **The Darkness of God: Theology after Hiroshima** (SCM 1982)

FOR REFLECTION AND DISCUSSION

1. SCIENCE AND MYSTERY

1. What role does the idea of purpose have in science?

2. What role does doubt play in religious belief?

3. Why is awe not a sufficient response to mystery?

4. Can there be certainty in religious matters? Explain your answer.

5. What is 'mystery' in a religious context?

6. What working assumptions does science make concerning the material world?

7. No one religious tradition has an exclusive claim on the truth. How essential is it that Christianity engage in inter-faith dialogue with other religions?

8. Since the whole world has virtually become one huge television audience, which of the traditional biblical titles for Christ is most likely to be universally understood? Explain.

2. THE SUN AND STARS

1. Is the God of the gaps finally dead? Explain.

2. What is life?

3. Does the new time scale of the universe as proposed by modern astrophysics postpone judgment Day indefinitely? Explain.

4. If the Big Bang model of creation is superseded by another theory, what effect would this have on theology?

5. Is creation a process or an event? Explain.

6. Can belief in God be based upon our 'reading' of the universe as claimed by natural theology, or does it depend ultimately upon divine revelation? Explain.

7. Is love more than a chemical phenomenon in this universe? Explain. What is it?

8. Is it true that 'The cosmos is all there is or ever was or ever will be'? (Carl Sagan) Explain.

3. GOD AND COSMOLOGY

1. God has created a universe in which we are free, free even not to believe in God. Does God believe in atheists? Explain.

2. If the world is governed by the laws of science, what difference does it make to have religious faith?

3. How do we identify revealed truth and distinguish it from speculation or theological daydreaming?

4. What do we mean when we refer to God having created the world?

5. How is the anthropic principal an argument for the Christian doctrine of creation?

6. Does the Second Law of Thermodynamics spell the ultimate death of God's creation? Explain.

7. By what authority does the reader interpret Scripture? Does the reader ever have authority to disagree with Scripture? If so, when?

8. Is it too anthropocentric to claim that the universe finds its fulfillment in the evolution of people? Explain.

9. What does it mean to say that the Bible is the 'Word of God'?

4. EVOLUTION

1. In what positive ways does the theory of evolution contribute to theological debate?

2. Where will evolution go next?

3. Does evolution have a direction? Explain.

4. Does God allow chance and accident to determine the course of evolution? Explain.

5. In what sense is 'creation science' a modern heresy?

6. Could the theory of evolution ever be disproved? Explain. What would constitute a disproof?

7. If you were a time traveler, how would you identify the historical moment when some primates had become people?

8. What are the various distinct or overlapping conceptual schemes that might be used to describe a human being?

5. THE HUMAN SPIRIT OR 'DOES MIND MATTER?'

1. Jesus as Messiah fulfilled the expectations of many of his contemporaries. What expectations does he fulfill today?

2. Mind has emerged through evolution as a product of complex chemistry. Are there any insuperable problems with the concept of a disembodied mind? If any, describe them.

3. Does the quantum theory and the indeterminacy principle in particular have any implications for the doctrine of free will? If so, what are they?

4. Will machines ever have 'souls' or become self-conscious? If not, then why not?

5. If Jesus is proclaimed as God's only Son, does this rule out the possibility that other incarnations have taken place elsewhere in the universe? Explain.

6. Has the human brain grown too large to be useful? Explain.

7. Jesus was a great human being. Was he more? Explain.

6. CATASTROPHES AND CANCERS

1. Does belief in God lead to belief in life after death? Why? Why not?

2. If at all, why would it matter if humankind joined the list of extinct species and evolution continued without us?

3. In God's eyes, is this world 'the best of all possible worlds'? Explain.

4. Does the doctrine of original sin have any place in an evolutionary account of creation? If so, what? If not, why?

5. Satan is dead; the new theology seems to have no use for the idea of evil personified. Is this wise? Explain.

6. The cross was inevitable in creation. How might this be true?

7. MIRACLES

1. Can we still expect miracles? How would you explain them?

2. Does prayer change things or does it only change people? Or neither? Explain.

3. Does God intervene in history? Explain.

4. Is the Christian gospel absolutely dependent on any historical facts? Explain.

5. 'I believe in the resurrection of the body' (Apostles Creed). How are we to understand this credal statement?

6. Is it a waste of time to pray for good weather? Why? Why not?

8. FAITH AND FALSIFICATION

1. Would you allow any experiment, argument, or piece of evidence to disprove your faith? What would it be? If not, then why not?

2. 'Except a man be born again, he cannot see the Kingdom of God' (John 3:3). How should a scientist approach this challenge to verification?

3. Is there any truth in astrology? Explain.

4. What theological statements can be thought of as having predictive value?

5. Does God exist outside of faith? Explain.

6. Does beauty have a place in scientific theory? What is it? If not, explain.

7. What role does imagination have to play in religious thinking?

8. Our nature demands for its fulfillment ends to aim at that lie outside ourselves. Is this true? Why? Why not?

9. PEOPLE AND THE ENVIRONMENT

1. Does the future exist in any way? Explain.

2. Are there any moral absolutes in an evolving world? Explain.

3. The doctrine of the incarnation encourages a sacramental view of the universe. Why do you agree, or disagree?

4. By what criteria could we say of an alien from space that he or she or it was made in the image of God?

5. Where in Scripture should we look for the roots of a theology of nature?

6. Ecology and a proper concern for the environment cannot proceed without theology. Why do you agree, or disagree?

7. Once intelligent life has appeared in the universe, it will never die out. Is this an unrealistic claim? Explain.

8. What are the consequences of acknowledging our kinship with the rest of the biosphere?

9. In what ways are biological facts relevant to ethical decision making?

INDEX